Maxime Collignon, John Henry Wright

A Manual of Greek Archaeology

Maxime Collignon, John Henry Wright

A Manual of Greek Archaeology

ISBN/EAN: 9783337120146

Printed in Europe, USA, Canada, Australia, Japan

Cover: Foto ©berggeist007 / pixelio.de

More available books at **www.hansebooks.com**

—◆◆◆—

MODERN culture owes to the civilisation of the ancient Greeks a profound debt, which is at once direct and indirect. The direct debt has arisen principally from the place long held by Greek studies in our system of education. The indirect debt, which is more subtle and less easily recognised, is that of many forces, inspirations, and models, in art, literature, and science, that have been transmitted to us from a remote past, through various peoples and through diverse civilisations. In our schools, and to a certain extent still in our colleges and universities, we understand by Greek studies almost exclusively the study of the language and literature of the ancient Hellenes. But the Hellenic spirit—and it is this only that gives life to these studies—has revealed itself in a novel and distinctly different manner, and with equal if not with greater vividness, delicacy, and force, in the manifold remains of Greek art, from the rudest specimens of the potter's industry, up to the glorious monuments due to the genius of the sculptor and architect in the out of view the art of the ancient Greeks, are one-sided, fragmentary, and essentially defective.

It is for reasons like the above that a cordial welcome should be extended, not only by lovers of Greek art, but also by students and teachers of Greek, to such a book as M. Collignon's handbook, which aims modestly to introduce the reader to these monuments of art, to "orient" him, as it were, both as to their general character and as to their historical relations and connections.

A few words on the scope and method of the book, in part taken from the author's preface to the French edition, are perhaps not out of place.

Forming one of a series of educational works on art (*Bibliothèque de l'Enseignement des Beaux-Arts*), it is above all an elementary text-book, designed for pupils in schools and colleges, and for amateurs in the study of art. The results of archæological research are as a rule recorded in elaborate scientific treatises or journals difficult of access to the average reader. These works, even when not in an unfamiliar language, are also commonly so technical in character as to disconcert and bewilder the beginner in his studies. Though there are, in some of the departments of classical archæology—as in sculpture, and in architecture—and even, for certain groups of departments, excellent handbooks of the nature of introductions, there seems to be no book in English, in line with recent research, which gives a bird's-eye view, as it

were, of the whole field, especia[l]   y
of Greek art.   The present book   ly
this deficiency.

After a brief *résumé* of th[e]   [n]e
beginnings of Greek art, and as [e]s
that moulded it more or less, its   [r]e
passed successively in review, the[   ]ns
being retained.   In each of these   [e]k
archæology, the monuments are   [o]-
logical order, selections from t[h]   [n]t
receiving special attention.   This   of
the several branches independent[   ]on
the reader the important truths t[h]   [n]i-
festations of the artistic spirit an[   ]re
manifold and varied, they are all s[   ]e
principles and to the same la[w]   [n]t.
Each illustrates the other, and   [o]r
the same native artistic genius.-[   ]al
classification has also the adva[n]   [n]g
with the classifications of wo[rk]   [a]rt
commonly adopted in our muse[um]   [o]l-
lections.   The book will thus a[   ] in
the museums in telling their own   and
consecutively.

The small size of the volume, as well as its aim,
precludes the introduction of any extensive or ex-
haustive scientific apparatus.   In the brief biblio-
graphies printed at the head of most chapters, a
selection only from the more important works on the

are the same in

conservatively

time assumed as they in

termined that

present-

angt of the

of Great art in the portions

ARCHÆ

Book First.

RIGIN OF GREEK

CHAPTER I.

GRÆCO-PELASGIC PERIOD.

Troy, 1875 ; *Ilios*, 1881 ; *Troja*, 1884.
ANT: *Les Antiquités de la Troade*, 1875.
Rapport sur une Mission scientifique à l'Île de Santorin : Archives des Missions scientifiques, Vol. IV., and *Santorin et ses Éruptions*, 1871.
Journal of Hellenic Studies, Vols. II., III.
LIEMANN: *Mycenæ*, 1878; *Tiryns*, 1885.
LENORMANT: *Les Antiquités de Mycènes : Gazette des Beaux-Arts*, February, April, 1879.
MILCHHÖFER: *Die Anfänge der Kunst in Griechenland*, 1883.
DUMONT ET CHAPLAIN: *Les Céramiques de la Grèce*, 1881, 1883 *ff.* (In course of publication.)

§ I. THE ANTIQUITIES OF HISSARLIK AND OF SANTORIN.

THE most ancient monuments left on the soil of Greece, both Asiatic and European, by her primitive inhabitants, date from a time when as yet Greece had no history. The empires of Egypt and Assyria had passed through long ages of prosperity before the early inhabitants of Greece had emerged from a low grade of civilisation. Occasional scattered references in documents written in Egypt, a few mythological

legends, and monuments recovered in successful excavations, are the only materials at the service of the
student of this obscure period.

Before the final establishment of the Dorians in
Peloponnesus, that great historical fact which closely
followed the Trojan War, we catch glimpses of a
long succession of migrations and of conflicts, the
theatre of which was in the countries bordering upon
the Ægean Sea. The great Aryan migration from
the East had in Asia Minor separated into three
groups. One, crossing the Hellespont and passing
through Macedonia, had settled down in the moun-
tainous regions of Macedonia and Thrace ; here arose
the Hellenic tribes that later descended into Hellas
proper. A second group had established itself on the
table-lands of Phrygia, whence it did not emerge. A
third finally occupied the coasts of Asia Minor, and
from thence colonised the islands of the Ægean and
a part of continental Greece. This is the Pelasgic
stock which the Greeks themselves regarded as
aboriginal, and whose monuments bear witness to
a high antiquity. Professor Curtius recognises in
them, not without reason, the ancestors of the
people whom he calls the Oriental Greeks :—" We
give to the maritime people of Asia Minor, to those
at least who belonged to the Phrygo-Pelasgic stock,
the name of Oriental Greeks." When the Hellenic
tribes, the Achæans, Dorians, Ionians, and Æolians
left Phthiotis (in Southern Thessaly), and spread over
the surface of Greece, they either drove from it, or
reduced to servitude, the Pelasgic inhabitants.

These Pelasgians, undoubtedly closely related to the Hellenes, appear in history long before them. Egyptian monuments, as early as the eighteenth dynasty, make mention of them, and in the reign of Seti and of Rameses II. (nineteenth dynasty) they are recorded as taking part in the expeditions made against Egypt by the Khetas (Hittites) of Syria and by the Libyans of Africa.

Little would be known of the state of the civilisation of this early people, if recent and most important excavations had not brought to light materials and documents that are entirely new. The discoveries at Hissarlik and at Santorin, those at Mycenæ and at Spata, reveal to us a civilisation that must have been common to the whole of the ancient Greek world. This Græco-Pelasgic civilisation extended to all the people occupying the basin of the Ægean Sea, who by means of an active coasting trade were continually in close relations with each other. It is impossible to fix, with perfect exactness, the date of the monuments discovered. These monuments, however, may be classified in two principal groups, that of Hissarlik and of Santorin, on the one hand, which carries us back to the very earliest times ; and, on the other hand, that of Mycenæ and of Spata, belonging to a more recent epoch, where Oriental influences begin to appear.

The objects found by Schliemann near the village of Hissarlik, in the Troad, belong, it seems, to the earliest civilisation. The important objects discovered here, which their discoverer would assign to the Homeric age, have given rise to much discussion. Schliemann exhumed the ruins of several superimposed cities. In the most ancient ruins, which showed evidence of a conflagration, the explorer believed he had found the traces of Homeric Ilium, and he gave the name of the Treasure of Priam to a rich collection of barbaric jewellery, containing vases of gold and of silver, beads of cast gold, etc., which were rescued from the débris. The identification of the ruins near Hissarlik with the Troy of Homer is far from being accepted without question. Certain scholars still place the city of Priam near Bunarbashi, following the opinion advocated in 1788 by Lechevalier. Hissarlik would thus mark the situation of Ilium of the Romans, or *Ilium recens*, which was often destroyed and rebuilt at the time of the Æolians, of the Lydians, of Lysimachus, and of the Cæsars.

Nevertheless, the view which places Homer's Troy at Hissarlik is very plausible. The beds of débris there accumulated, to the depth of sixteen metres, prove that for ages a dense population had inhabited the Hill of Hissarlik ; furthermore, this place is nearer the sea than Bunarbashi, and thus is more in harmony with the scenes and situations presented in Homer.*

As to the objects found at Hissarlik, it is impossible to see in them, as would Dr. Schliemann, the remains of the civilisation described in Homer. They

* Compare G. Perrot, *Les Découvertes du Dr. Schliemann Troie et Mycènes*, a lecture at the Sorbonne, March 19, 1881.

belong to a semi-barbaric age, and the people that made them hardly knew the use of the metals. No trace of Egyptian or of Babylonian influence can be detected in them, nor again anything truly Hellenic. The pottery, in particular, is entirely hand-moulded, and is very primitive. In its technique it recalls the pottery found beneath the tufa in ancient Latium, and the vases of Santorin anterior to the eruption. The subjects with which it is decorated are childish representations of animals, and the vases themselves, in their outlines, rudely imitate the human figure.

Before the sixteenth century B.C. (about B.C. 2000, according to Fouqué), the island of Thera, one of the Cyclades, was disturbed by a volcanic eruption ; the central portion sank, leaving only a circular fringe formed by the islands now known as Santorin, Therasia, and Aspronisi. Excavations carried on at Therasia and at Santorin in 1866—7, brought to light the ruins of habitations built before the eruption. In the midst of the débris of rudely constructed houses, under a mass of pumiceous tufa, were recovered many objects used in daily life, hand-mills, little troughs, etc., and, in particular, vases of primitive ——— style, with swelling necks imitating female forms. Objects in the same style, recovered above the bed of tufa, proved that, a short time after the eruption, another population had taken the place of the earlier. This latter people, who inhabited the island before its occupation in the fifteenth century by the Phœnicians, lived by fishing and cultivating the ground. Perhaps the ancient colonists of Thera may be recognised in an Egyptian painting on the tomb of Thothmes III., at Rekhmara, where "the people of the isles of the sea" bring gifts to Pharaoh; in their hands they hold vases with tilted mouths, which by their shape recall the vases of Santorin.

Together with the antiquities of Hissarlik, those of Santorin constitute, in the present state of archæological science, the most ancient group. They are the earliest materials preserved to us for the study of industrial art among the ancient Greeks.*

§ 2. THE ANTIQUITIES OF MYCENÆ, OF SPATA, AND OF RHODES.

The second group seems to belong to a more recent epoch, and to represent a more advanced phase of civilisation, the beginnings of which are discernible at Hissarlik and Santorin. The acropolis of Mycenæ had long attracted the attention of scholars by its sculptures of the "Gate of Lions," and by its ruins of gigantic walls. Here, in 1874, Dr. Schliemann began a series of excavations, which led to the discovery of five tombs, containing rich funeral equipments. The discoveries at Mycenæ are a veritable treasure. Numerous objects in gold, including plaques, or small flat plates, done in repoussé, crowns, cups, funeral masks, imitating the human visage, and placed on the faces of the dead ; arms, objects in bronze, crystal,

---

* No mention is here made of the monuments belonging to the age of stone, which have been found in Greece, as in the whole of Europe. Compare Dumont, *La Grèce avant la Légende et avant l'Histoire: Revue archéologique*, Vol. XVI.

alabaster, engraved stones, fragments of *stelæ*, make up this treasure, which comprises at least twenty

FIG. I.—FUNERAL MASK IN GOLD.
(Found at Mycenæ.)

thousand separate pieces. There is but little iron ; the metals most frequently used are gold, copper, and bronze.

Some of the objects evidently show that they were imported from the East; such is the model of the temple with doves, of which a drawing is given below (Fig. 2); such also are some beautiful golden vases,

FIG. 2.—MODEL OF A TEMPLE IN GOLD.
(Found at Mycenæ.)

decorated with radiated flowers, and lions of a conventional type.

Oriental influence is likewise detected in the golden stamped plaques and the *bracteæ* or leaves of metal, which are sewn upon garments. But while objects of Phœnician and Babylonian origin, imported by commerce, may be distinguished among the treasures from Mycenæ, the greater part of these treasures are the product of local industry,

FIG. 3.—GOLDEN STUD IN REPOUSSÉ.
(Found at Mycenæ.)

and are still rude and imperfect in style. Among these are the gold vases, a large breastplate of the same material, and gold studs, hammered and cut, which decorated objects of wood or leather, such as the scabbards of swords.

These remains exhibit a very original system of ornamentation, in which curves and bent lines are almost exclusively employed. Such motives as the following constantly occur : spirals, floral rosettes, circular bosses decorated by points in relief or by concentric circles, the foliage of aquatic plants, imitations of insects and of marine animals, as polypi, medusæ, star-fish, etc. This same system prevails in the pottery, made by the wheel and decorated with paintings, found in tombs ; hence the local origin of these vases can no longer be in doubt.

It is difficult to believe, with Schliemann, that the tombs discovered are actually those of Agamemnon and his companions, murdered by Ægisthus and Clytæmnestra. It is nevertheless probable that these monuments date from the epoch of Achæan domination in Peloponnesus. Ottfried Müller has proved that in the heroic age the Achæan race occupied the greater part of Peloponnesus and Thessaly, and this race is perhaps named on Egyptian monuments of the fourteenth century B.C. When, under the reign of Merenptah, the Mediterranean nations made attempts against Egypt, the Achæans seem to have taken part in them ; their name is possibly recognisable in "Akaios" in an inscription at Karnak, beside the supposed Tyrrheni, the Siculi, the Sardinians, and the Lycians or "Leka." The civilisation of this people exhibits a barbaric grandeur ; gold is lavishly used in the sepulchres of the Achæan chiefs at Mycenæ. But the Homeric age is still in the distance ; two centuries at least separate the art of Mycenæ from that described by the Homeric poets. The objects made by the rude and primitive goldsmith of Mycenæ resemble much more the treasures dis-

covered in the barbaric tombs in the Danube valley, especially at Hallstadt, near Vienna, than the works of Greek art of the subsequent period. The Achæans of Mycenæ were already in commercial relation with the Phœnicians, but the time had not yet arrived when the East was to initiate Greece into the plastic arts.

The antiquities of Mycenæ are by no means an isolated fact. Unique as they appear at the first glance, they receive fresh light from discoveries made in different parts of the Greek world, and above all from the excavations at Spata, a small village in Attica.*

FIG. 4. — SPECIMEN OF AP-
PLIED WORK IN IVORY.
(Found at Spata.)

Glass-ware, articles in gold and ivory, found in the subterranean tombs of Spata, testify to an artistic industry more advanced than appears at Mycenæ. Oriental influence is here much more apparent. The ivory head from Spata of a bearded man

* Compare the list of objects discovered at Spata by Haussoullier, *Bulletin de Correspondance hellénique*, Vol. II.

wearing a conical mitre recalls very closely certain
Phœnician statues found in Cyprus, and a design
represented on an ivory plaque from Spata — a
lion devouring an ox — is not unlike one which
occurs both on an engraved stone of Phœnico-
Sardinian origin, and on a Phœnician bowl from
Palestrina.   But while Oriental influence is here
more marked than at Mycenæ, products of a local
industry are also to be recognised, decorated with
designs suggested by the imitation of aquatic vege-
tation, or of marine birds, and of fishes, such as
characterises all the Græco-Pelasgic ornamentation.

   To these discoveries have lately been added others
at Ialysus on the island of Rhodes, in Cyprus, of the
archaic tombs at Nauplia in Argolis, and at Menidhi
in Attica.   On the strength of these discoveries we may
form some idea of the civilisation common to all the
inhabitants of the Oriental shores of the Mediterranean
before the return of the Heracleidæ, the epochs of
which are clearly defined.   Dumont indicates the
following as approximate dates : " before the sixteenth century, Hissarlik ; sixteenth century, Santorin ; fourteenth, Ialysus ; thirteenth, or twelfth,
Mycenæ and Spata."[*]

   We thus approach the Græco-Oriental period,
which succeeds the previous civilisation by a very
natural transition.   This is the time when Greek
genius is becoming thoroughly impregnated with
Oriental influences, but is still feeling its way.   The

   [*] A. Dumont et Chaplain, *Les Céramiques de la Grèce propre*,
1881, Ier fascic., p. 75.

rading settlements or *emporia* of the Phœnicians, ɔcated among the islands of the Ægean and along he shores of Peloponnesus, imported types of work-ɲanship which the Greek with a naïve spirit of rtistic imitation endeavoured to reproduce. On the ther hand Ionian Greece, already half Oriental, ɔrdered upon Lydia, and came into contact with the eoples of Asia Minor, whose art was distinctly and ɲoroughly under the control of Assyrian influences. ïreek art awoke slowly and with difficulty, after assing through a long period of imitation, which ɔntinued until towards the close of the seventh entury B.C. It is important, therefore, that we ɲould examine the question as to what it owed to ɲe more advanced civilisations which gave it its rst models for imitation.

CHAPTER II.

ORIENTAL SOURCES OF GREEK ART.

GERHARD: *Ueber die Kunst der Phönicier* (*Gesammelte Akad. Abhandlungen* 1867—8).

RENAN: *Mission de Phénicie*, 1864—1874.

CESNOLA: *Cyprus*, 1878 ; *Atlas of the Cesnola Collection*, Vol. I., 1884.

DE CHABAS: *L'Antiquité historique et les Monuments égyptiens,* 1873.

LEPSIUS: *Ueber einige ægyptische Kunst-formen und ihre Entwickelung* (*Abh. Kön. Preuss. Akad. d. Wiss. zu Berlin*).

DE LONGPÉRIER: *Musée Napoléon III.*, 1865—1874.

LAYARD: *Nineveh and its Remains*, 1850 ; and *The Monuments of Nineveh*, 1849

PLACE: *Ninive et l'Assyrie*, 1865.

PERROT and GUILLAUME: *Exploration archéologique de la Galatie et de l Bithynie*, 1872.

PERROT and CHIPIEZ: *Histoire de l'Art dans l'Antiquité*, 1882 *ff.* (translated b Armstrong, *Egypt*, 1883 ; *Chaldæa and Assyria*, 1884 ; *Phœnicia an Cyprus*, 1885).

PERROT: *L'Art de l'Asie Mineure* (*Mélanges d'Archéologie*, 1875).

SEMPER: *Der Stil in den technischen und tektonischen Künsten* (1860-3) 1878-9.

HEUZEY: *Catalogue des Figurines antiques de Terre-cuite du Louvre*, 1878.

THE Greeks seem to have attempted to rende the origin of their art obscure. If we are to believ what they say of themselves, they invented everything and the earlier writers on the history of Greek art knowing but little of the East, have given currency to this erroneous opinion. "Art," says Winckel mann,[*] "though born much later among the Greek than among the Orientals, began there with the humblest elements, and it exhibits a simplicity which easily convinces us that the Greeks took nothing

---

[*] *History of Art*, chap. i. §§ 7, 8.

from the art of other nations, but invented their own art."

It has, however, been thoroughly proved since Winckelmann's time, that Greek art at the outset was subject to the influences of Oriental civilisations, that it owed to them its first models, and that it received from them a knowledge of methods of execution as well as technical skill.

A glance, moreover, at the geographical situation of Greece suffices to show us how favourably she was placed in respect to a continuous commerce with the East ; the islands scattered in the Ægean Sea, at distances apart so slight that they could be traversed in a few hours, put Greece into direct communication with Egypt and Phœnicia. Again, the great valleys of Asia Minor were equally natural routes of travel opened up in the direction of Assyria. These material conditions, and the inferiority of Greece, in the midst of the advanced and flourishing civilisations of older peoples, contributed to make her the pupil of the East.

§ I. PHŒNICIAN INFLUENCES.

It was upon the Greeks of the islands and of Peloponnesus that Phœnician industries exercised the most potent influence. The colonies of the Sidonians and of the Tyrians had made of the Mediterranean a Phœnician sea : their trading settlements and factories had been established at Rhodes, on Crete, on the Cyclades, and as far west as Cythera. From Cythera the Phœnicians passed over into Peloponnesus up to Amyclæ and Gythium, and advanced as far even as Argolis, Attica, and Bœotia. These establishments placed them in close contact with the occidental Greeks, above all with the Dorians. What Greek civilisation owed to these Phœnicians is generally recognised ; above all the alphabet.

Phœnician ships brought into Greece works of gold, silver, glass, and ivory, manufactured by the glass-workers and goldsmiths of Tyre and Sidon ; painted vases, statuettes of bronze and of terra-cotta, like those that were sold at Paphos in the seventh century B.C., and that served the sailors as talismans. These objects became the models for Greek workmen, who imitated them with childish awkwardness. Thus a whole class of painted vases discovered in the Cyclades shows how, in the twelfth and thirteenth centuries B.C., Greek potters copied the products of Phœnician art. The Sidonian and Tyrian navigators imported more than the products of their own industry.

By a privilege acquired in the reign of Thothmes I., they had received a monopoly of Egyptian commerce with foreign lands, and they had thus scattered over Greece articles of Egyptian manufacture which, because Egypt was less known to them, must have made a marked impression upon the inhabitants. To the Phœnicians then must be ascribed the double rôle, that of intermediaries between Greece and Egypt, and that of initiators in respect of their own peculiar industries.

To what degree did Phœnician art act upon the

ıascent art of Greece ?　It offered, for the imitation of
he Greeks, works of extremely mixed styles.　It has
ıften been noticed that Phœnician productions never
lo more than reflect the style of Egypt or Assyria,
ıccording as each nation had the political supremacy
ıver Phœnicia.　The most conclusive proof of this is
urnished by the monuments of Cypriote art, one of
he most important branches of Phœnician art.　In
he discoveries of General Palma di Cesnola in the
sland of Cyprus we possess a rich series of statues
:oming from the ancient cities of Golgos and of
:dalium.　In them may be found predominant suc-
:essively the influences of Egypt and of Assyria.　The
:tatues in Egyptian style show erect figures with
ırms hanging parallel to the body, with the *klaft* or
Egyptian *pshent* for head-dress ; about their loins are
ɪorn the *shenti* or sloping short trousers.　Other
:tatues, in the Assyrian style, contrast with the pre-
:eding ; in these the figures, kings or priests, wear a
:ind of pointed cap ;* the beard and hair are arranged
ɪ symmetrical coils, and a long robe covers almost
:he entire body.　Finally, statues of a later date
:uggest the style of archaic Greek art, but all have a
:ommon basis or family resemblance, which con-
:titutes the Cypriote type.　These successive changes,
lue to changes in foreign influences, can be followed
ɪ the valuable collection of *figurines* in the Louvre,
he fruit of the excavations of M. de Vogüé.　In these

---

* *The Antiquities of Cyprus*, edited by Newton and Sidney
:olvin, 1873, plates IX., XVIII. ; Perrot and Chipiez, *Art in Phœnicia
·nd Cyprus*.

C

can be recognised the styles of Egypt and of Assyria down to the time when Cyprus had become one of the centres of Greek civilisation in the East.

It is not difficult to grasp the style of the models offered by Phœnicia to the Greeks for imitation ; in general, we find the Egyptian form, with an Assyrian carefulness as to detail and execution. This combination can be seen in the important monuments discovered in Cyprus. The collection known as the Curium treasure comprises numerous objects in which the Egyptian and Assyrian styles appear at the same time ; to the former belong the scarabs and gilded cups ; to the latter the clasps decorated with chimæras and flowers, as well as the cups ornamented with subjects familiar to the artists of the kingdom of Assur. Beautiful silver-gilt cups of Phœnician workmanship, found at Larnaca, present the same characteristics ; the attitude and the costumes of the figures represented on the friezes, and the details of ornamentation, show such a confusion of styles, that we may recognise at one and the same time the *uræus* of the Egyptian kings and motives employed in the decoration of the palaces of Nineveh.*

The influence of Phœnicia upon the industrial art of Greece is incontestable ; in the domain of sculpture however, this influence is less distinct. Phœnicia did not possess a style sufficiently original and characteristic to impress itself upon the earliest Greek sculptors. The researches of Heuzey have shown that Phœnicia

* De Longpérier, *Musée Napoléon III.*, plates X. and XI.

on the contrary, early subjected her art to the influence of the archaic art of the Greeks, as it was developed in the sixth century B.C. in the Greek cities of Asia Minor and the islands.*

§ 2. EGYPTIAN INFLUENCES.

The direct influence of Egypt upon the art of Greece has been greatly exaggerated. It is certain that for a long time Egypt was closed to the Greeks, and known to them only through the medium of the Phœnicians. It was not opened to them until the twenty-sixth dynasty (Saïtic), in the reign of Psammetichus I. (seventh century B.C.), and at that time the Greeks were already in possession of the technical processes of art. The historians tell us of the astonishment with which Egyptian civilisation struck the Greeks. The influence of Egypt, however, was somewhat felt at the beginnings of Greek art. Pausanias, speaking of the ancient ξόανα, the primitive images of Greek divinities, declares that many of them were Egyptian. According to him the wooden statues of Heracles, of Hermes, and of Theseus, in the gymnasium at Messene, clearly show Egyptian origin, and the same style may be exactly (ἀκριβῶς) recognised in the Heracles of Erythræ, brought from Tyre by the Phœnicians.†

Pausanias divides the ξόανα into two classes—those wrought in the Egyptian style, or brought from Egypt, and those of the Dædalidæ, the pupils of

---

* Heuzey, *Catalogue des Figurines antiques du Louvre*, 1882.
† Pausanias, VII. 5.

C 2

Dædalus ; in other words, in the view of the Greeks, their most ancient religious statues follow the Egyptian tradition, and in Dædalus are to be seen the first attempts of Greek art to emancipate itself from this tradition.

We have noted the part played by Phœnicia in this matter of the Egyptian origin of Greek art. Through her commerce she imported into Greece objects that served for models ; by the Egyptian character of her art, she gave to the early examples of Greek art a reflection, as it were, of the arts of Egypt. When Psammetichus, by subduing the Ionian and Carian pirates in the seventh century B.C., opened his kingdom to the Greeks, Hellenic genius had begun to emerge from its long infancy ; it was then ready to receive from Egypt that which it actually seems to have borrowed—a profounder and more religious feeling in art. In architecture the budding Doric was inspired afresh by the massive forms of the Egyptian column ; in sculpture, the Greek artists, following the Egyptian, applied to the human figure the principle of a more exact canon. This Egyptian influence may be detected in many archaic Greek sculptures. A statue of Artemis, found at Delos, and made by a Naxian* in the seventh century B.C., is a reproduction of those Egyptian statues of wood (ξόανα Αἰγύπτια), of which Pausanias spoke ; the arms hang close to the body, and the legs seem encased in a sort of sheath. Imitation of Egyptian art is no less

* _Bulletin de Correspondance hellénique_, Vol. III., plate I. This statue was discovered at Delos by M. Homolle.

visible in a statue of a lioness of calcareous stone
(Fig. 5), of a date subsequent to that of the Delian
Artemis described above.

§ 3. ASSYRIAN INFLUENCES.

The part played by Assyria in the history of the
Oriental sources of Greek art is of the utmost importance. This influence was shown, above all, in Asiatic

FIG. 5.—STATUE OF A LIONESS.
(Found at Corfu.)

Greece, in Ionia, where art was first developed, and in
some parts of Greece proper—at Corinth for example,
a commercial city, the business relations of which
placed it in direct connection with Asia Minor. The
discoveries made at Nineveh by Botta, and the
excavations of Layard at Koyunjik and at Nimrûd,
are of the highest importance as throwing light on
the history of Greek art. The comparison of
Assyrian monuments with the most ancient
Hellenic works has shown clearly that in Ionia
Hellenic art was formed in the school of Assyria.

This relationship may be proved from a large number of facts, which may be grouped as follows:—
(1) motives in ornamentation; (2) types of the human figure and of animals; and (3) technique, both ornamental and plastic.

(1) Certain motives, or subjects of decoration, have passed directly from the *stelæ* and enamelled bricks of Assyria to the painted vases and marbles of Greece. Such are the palm-leaf ornament and the rosette which appear upon Greek vases of the most ancient style. The lotus flower, in full bloom between two buds, is Assyrian, and is often met with in Greece in ceramic paintings of the Corinthian style.

(2) This imitation is no less visible in subjects comprising types of animals or of the human figure. It is the East which has created all that fantastic world of sphinxes, of winged figures, of impossible animals with human heads, those belts of tigers, rams, and moufflons, which march in long files on bas-reliefs, or on the surface of the metal cups of Nineveh, and also find a place on archaic Greek vases. The resemblance between the Assyrian griffins found at Nimrûd, and those that decorate the vases of Rhodes, is equally striking. These monstrous and fantastic figures of the Assyrians were frequent in early Greek art, but no longer appear on later monuments. The earliest Greek artists did not limit themselves to the imitation of Assyrian subjects; they also copied the Assyrian system of decoration. Their vases were ornamented, as were the metal cups of Cyprus and of

Nineveh, with successive zones, which resembled so
many superimposed friezes.   The bronze *crater*, which
the Dorians of Sparta had ordered for Crœsus, was
ornamented in the same way.   "It was," says Herodotus, "decorated up to the brim with the figures of
plants and of animals."*

Besides the vases and articles in metal, carpets
and other rich stuffs of Assyria furnished the Greek
with these types of ornamentation.   Thus it was that
the peplos of Alcimenes of Sybaris had its border
decorated, as· Aristotle tells us, with Oriental subjects.   "The upper portion represented the sacred
animals of the Susii, and the lower those of the
Persians."†

(3) Assyrian influence betrayed itself likewise in the
earlier examples of Greek sculpture. Even though, from
the beginning, the Greeks showed themselves much
more original in the plastic than in the industrial arts,
such as pottery, there was here also a distinct debt to
the East.   The carefulness of detail, the attention
with which the accessories of beard, hair, and costume
are treated, a certain tendency to accent the anatomy
in the nude, causing the muscles to stand out, heavy
and thickset figures—are features common alike to
the plastic art of Assyria and to that of these earlier
Greeks.   It should be admitted, however, that in this
field of art, imitation is more difficult, and that direct
study of the nude enabled the Greeks to develop their
individual peculiarities with much greater rapidity.

* Herodotus, I. 70.
† Aristotle, *Mir. Ausc.*, 96.

§ 4. LYDO-PHRYGIAN ART.

The dominion of Assyria in Cyprus and in Phœnicia at the time of the Sargons is not sufficient to explain the influences of which we have just spoken. The transmission to Greece of Assyrian forms and technique was effected chiefly through Asia Minor ; we are able to study chiefly the art serving as intermediary through the researches of G. Perrot. In Pteria and in Phrygia, at Euyuk, at Boghaz-Kieui, at Kalaba, this Lydo-Phrygian art can best be studied. It seems to have been common to Lydia, Cappadocia, and Phrygia, and is directly connected with Assyria. In the figures of animals—lions and bulls—may be recognised an exact imitation of the Assyrian types which commerce had scattered over Asia Minor. On seeing the bas-relief of Pteria, where figures in hieratic attitudes clothed in Oriental costumes advance in long files, it is difficult not to be reminded of the sculptures of Nineveh. Doubt as to the rôle played by Asia Minor in the beginnings of Greek art is no longer admissible.* Recent discoveries have only confirmed

---

* ["HITTITE" ART.—Among the many monuments of inland Asia Minor, here called Lydo-Phrygian, is an important series exhibiting a peculiar art, in which the originals of several types in Phrygian art, strictly so called, may be detected. These monuments, which are found principally in Pteria (Cappadocia), but also in Phrygia, and at Karabel (the "Sesostris" of Herodotus), and Mt. Sipylus (the so-called "Niobe"), in Lydia, commonly represent figures in relief, in stiff hieratic attitudes, wearing shoes with ends turned up, and either a pointed or a cylindrical shaped cap (polus) ; inscriptions, which have not yet been deciphered, of a peculiar character, written boustrophedon,

the theory of Gerhard, who, speaking of subjects borrowed by primitive Greek art from Assyria, remarks : "These artistic types seem to have been imported into Greece less by the Phœnicians than by the people of Asia Minor, who had control of the commercial routes that pass through Comana and Tarsus and finally terminate at Nineveh and Babylon." *

frequently accompanying the sculptures. The art represented by these monuments was undoubtedly the most important medium through which the art of Mesopotamia, more or less modified in the process, was transmitted overland to the Phrygians, and through them to the people dwelling on the coasts of the Ægean.

These monuments, both in style and in some of the hieroglyphic characters used in the accompanying inscriptions, bear a striking, though in some respects an illusory, resemblance to certain monuments found in Northern Syria, ascribed to the ancient Hittites ("Kheta" of the Karnak inscription of Rameses II., the Hittites of the A.V. of the Bible). On the strength of these resemblances, some scholars would ascribe these monuments to the civilisation and art of the Hittites, who, according to Sayce, the most prominent advocate of this theory, were a non-Aryan people with a kingdom extending about the fourteenth century B.C. as far west as the Ægean. (Sayce, *Trans. Soc. Biblical Arch.*, Vol. VII. ; *Herodotos*, 1883, pp. 425 *ff.*; *Academy*, Aug. 18, 1883 ; W. Wright, *Empire of the Hittites*, 1884 ; Ebers, *Annali dell' Inst.*, 1883, p. 109).

The arguments adduced in favour of this theory, and in fact of the Hittite conquest of central and western Asia Minor, cannot be regarded as conclusive, and the Hittite origin of these monuments is accepted by but few archæologists. For this early, and as yet enigmatical, phase of art in Asia Minor, the designation "Anatolian" has been proposed. *Cf.* Ramsay, *Athenæum*, Dec. 27, 1884 ; *Journal of Hellenic Studies*, Vols. III. *ff.* ; G. Hirschfeld, *Paphlagonische Felsengräber* (*Abh. Berliner Akad.*), 1885. Other literature on the Hittites in Reinach, *Manuel de Philologie classique*, Vol. II., p. 77—78. Compare Perrot et Chipiez, *Histoire de l'Art dans l'Antiquité*, Vol. III. (Phenicé, Chypre, Asie Mineure), 1884.]

* *Ueber die Kunst der Phönicier.*

Such are the data which modern archæological science has substituted for the fables wherein the Greeks obscured the sources and beginnings of their art. The Lydian Cyclopes, the Dactyli of Mount Ida, the first skilled workers in iron and other metals, and the Telchines coming from Crete to Cyprus and Rhodes, all symbolise a fabulous art. But even beneath these legends we may discern what the Greek dimly recognised as to the forms and processes of art transmitted from the East to Greece. Greek art has thus followed a natural law : the latest comer, though with some original traits which are apparent in its first attempts, it has borrowed from anterior civilisations all that could be learned, and finally, after a vigorous effort, has brought into independent existence its own original qualities.

CHAPTER III.

THE GRÆCO-ORIENTAL PERIOD.

BRUNN: *Die Kunst bei Homer,* 1868.
HELBIG: *Das homerische Epos aus den Denkmälern erläutert,* 1884.
CONZE: *Melische Thongefässe,* 1862 *f.*
DUMONT ET CHAPLAIN: *Les Céramiques de la Grèce propre,* IIe fascic., 1883.
SALZMANN: *Nécropole de Camiros,* 1867—73.

§ I. ART IN THE HOMERIC AGE.

WE do not venture to fix accurately the date when the influences of the East upon Greek art began to be felt. In matters of this sort too positive affirmations are a source of error. At the same time it is safe to say that, from the end of the seventh century B.C., the Greeks were in full possession of technical processes, and that at about that date the history of Greek art really begins. The period immediately preceding this time, which we have named the Græco-Oriental period, is marked by the efforts of Greek genius to emancipate itself and to overcome the influences to which it had necessarily been subject.

Monuments of this period are rare. Our knowledge of Greek civilisation, from the Trojan War and the Dorian invasion down to historic times, is largely to be gathered from the texts of the Greek authors.

These texts themselves show how much Greece owed
to the East. The Homeric poems, the date of which
must be placed nearer the ninth century B.C. than
the Trojan War, describe the civilisation of those
times. They ascribe to the personages of the heroic
age contemporaneous manners and customs: Homeric
civilisation is half Oriental. The buildings described
in the *Odyssey* show an architecture of an Assyrian
rather than of a Greek type. The palace of Alcinoüs
is an Oriental palace ; brilliant colours, precious
metals, are there scattered in rich profusion, and give
it " a splendour like that of the sun or of the moon."
The walls of bronze may be explained by the plaques
of beaten bronze which adorned the palaces of
Assyria ; walls brilliant with a dark blue colour (περὶ
δὲ θριγκὸς κυάνοιο)* call to mind the enamelled bricks
of Nineveh. In the dogs of gold and silver, fashioned
by Hephæstus, which guarded the portals, we may
recognise the counterpart of those fabulous animals,
winged bulls with human faces, that stood at the gates
of the palace of Khorsabad. The works of art described by Homer show that the most advanced art
was that of working in metals, but the poems show
no knowledge of the use of solder, subsequently invented by a Greek of Chios. The complicated shield
of Achilles was covered with small figures of gold and
of silver, hammered (σφυρήλατα) and put together
mechanically : the figures were without doubt arranged in zones or belts. This was art in the style of

* Hom. *Odyss.* vii. 87.

the Ionian Greeks of the tenth century B.C., pupils of the Assyrians and Phœnicians. Homer speaks, to be sure, of vases of great value,* but these craters were, in the words of the poet, the work of Sidonians.

§ 2. ART IN THE SEVENTH CENTURY B.C.

In the seventh century B.C., a time at last historic, the texts show us that the mixture of Asiatic and Hellenic influences is still characteristic of the period. Pausanias has left a description of an important monument of the seventh century B.C., the chest of Cypselus, dedicated at Olympia by the Cypselidæ in memory of Cypselus, tyrant of Corinth, whom his mother had concealed in a chest in order to save his life.† Current opinion placed this monument at about the thirtieth Olympiad.‡

It was decorated in horizontal belts, some of the figures being carved in the cedar, while others, of gold or of ivory, were inlaid. The subjects represented were taken in great part from the Hellenic myths, but the influence of the East is still deeply felt in their treatment. This influence betrays itself by the striving after a crude symbolism and by the frightful character of some of the figures, such as that, for instance, of Destiny (Κήρ), represented with the features of a woman, but with hooked nails and enormous teeth.

---

* Hom. *Il.*, xxiii. 740, *Odyss.* iv. 616.

† Pausanias, v. 17—19.

‡ It may have been made much earlier. Pausanias says only that the inscriptions accompanying the subjects were taken from the poet Eumelus, who flourished towards the close of the ninth Olympiad (741 B.C.).

Certain subjects purely Oriental were not understood
by Pausanias, such as the Persian Artemis.　This Greek
traveller asks why she is represented "with wings on
her shoulders, holding a panther in one hand and a

FIG. 6.—PERSIAN ARTEMIS.
(From a Greek vase.)

lion in the other," a design often reproduced on
Oriental gems and on the Phœnico-Greek ornaments
from Cameirus in Rhodes.

　　An idea as to the style of the figures may be obtained from the vase paintings of the eighth and
seventh centuries B.C.　We have already spoken of the
vases of Rhodes and of Corinth; those of Melos are

no less interesting. The ornaments are still Oriental ; belts of Asiatic animals are still found, but the figures they mark off and include are now Hellenic gods in a Greek form. On a vase from Melos may be seen Apollo and Artemis treated as the figures on the chest of Cypselus must have been.

Soon after the second half of the seventh century B.C., schools of art were instituted in Oriental Greece, and the art of working in metal underwent a remarkable development. Eastern art is no longer slavishly copied. Greek art, as it were, realises itself, and becomes something individual. About the fortieth Olympiad (or about the twentieth, according to the chronicle of Eusebius) Glaucus of Chios invented the art of soldering metals ; thus was substituted a new process for the old one of putting pieces together mechanically. This old method was employed before the thirty-eighth Olympiad, when a colossal figure, designed for Olympia, made of sheets of hammered gold, riveted with nails, was executed for the Cypselidæ. At Chios likewise, the sculptors Melas, Micciades, and Archermus, in the seventh century B.C., were the founders of a school which developed brilliantly in the sixth century B.C. At Samos the art of working in bronze made rapid progress under the impulse given it by Rhœcus and his sons Theodorus and Telecles. These workers in metal (τορευτάι) were also architects. They began the great temple of Hera (Heræum), the construction of which demanded

---

* Conze, *Mel. Thongefässe*, plate IV.

a large variety in workmanship, and displayed the many-sided talents of these old masters. In the seventh century B.C., this Samian school of bronze-casters produced some important works, such as the bronze crater dedicated in the Heræum by the Samians on their return from Tartessus (in the thirty-seventh Olympiad). This crater was orna-mented with griffins' heads in round bosses, three kneeling figures serving as a pedestal. The artists of Samos acquired such skill that their works were in demand in the East less than a century later. This Samian school also executed for Crœsus a crater of gold, which many years afterwards was in use in the palaces of the Persian kings (Olymp. LV.—LVIII.).

This rapid progress in art took place during this period almost entirely among the Oriental Greeks. While in the seventh century B.C. many temples were built in these regions (at Samos, Sardis, and Ephesus), the Dorian districts of European Greece could count up but a small number. Before long, however, the de-velopment of art proceeded equally among the western Greeks as among the eastern Greeks. About the fortieth Olympiad the Dorian schools blossomed into life under the influence of the Cretan sculptors Dipœnus and Scyllis, and of the Magnesian Bathycles. The orders of architecture were formed ; to the ancient rudely shaped wooden images succeeded statues of gods and heroes that bore clear testimony to a direct study of nature ; sculptors ceased to be called "stone-cutters," as the first artists working in marble were termed. A hundred years, however, still

separated Greek art from the wonderful fifth century
B.C., that epoch of its perfection.

While the traces of Oriental influence grow fainter,
and in time become actually effaced, the contrary
tendencies of Doric and Ionic genius become more
and more marked. In spite, however, of these differences, there is still in all this early art a character
common to the whole Hellenic race ; a fine instinct
for the beautiful, a supreme faith in its own genius, a
disdain for everything un-Hellenic.

D

Book Second.

ARCHITECTURE.

CHAPTER I.

GRÆCO-PELASGIC MONUMENTS.

MIDDLETON : *Grecian Remains in Italy, a Description of Cyclopian Walls, etc.*
1812.
PETIT RADEL : *Recherches sur les Monuments cyclopéens, etc.*, 1841.
DODWELL : *Views and Descriptions of Cyclopean or Pelasgic Remains*, 1834.

WE shall give only slight attention to monuments anterior to the appearance of the Orders in architecture. Art can have but little to do with the erection of those massive structures raised by the Pelasgians in Asia Minor and Italy, as well as in Greece. The Greeks, struck with astonishment at them, believed them to be of mythological origin; they ascribed them to fabulous beings, the Cyclopes or the *Gasterocheires* of Lycia. The ruins, known indiscriminately at the present day as Pelasgic or Cyclopean, belong to diverse epochs, and are to be classified in accordance with the differences shown in the structure of their walls.

The most ancient are commonly termed *Cyclopean walls*. These ramparts are formed of enormous blocks, put together without cement, with smaller stones filling up the interstices. The most striking

D 2

examples of this method of construction are found in Argolis, in the corridor-like galleries of Tiryns. These galleries, built between thick Cyclopean walls, end in narrow triangular doors, and were planned with a view to defence.

The constructions called *Pelasgic* are composed of huge blocks, executed with greater regularity ; these blocks are polygonal in shape, well fitted together, and finished with smooth outer surfaces. This type is met with in several parts of Italy and Greece ; a portion of the walls of Mycenæ exhibits it.

The Pelasgic construction includes a second variety, which has sometimes been termed the *third polygonal system*. In this the blocks begin to assume a quadrangular form, but the layers are not horizontal, and the lines of juncture cross in every direction. This construction was employed at Mycenæ in that portion of the walls of the Acropolis adjoining the Gate of Lions. There is reason to believe that these walls are of a more recent period than the Cyclopean masonry, and that they belong to the Achæan age. At any rate, we cannot give them a precise date. Euripides is only echoing popular tradition when he ascribes them to the Cyclopes, who built them "with lever, rule, and hammer."*

These massive walls clearly enough declare that the chief concern in their construction by the ancient inhabitants of Greece was the provision for defence in case of attack. Cities were merely places of

---

* Euripides, *Herc. fur.* 943 *f.*

refuge, built upon high hills.   In time of alarm everything that could be saved was hurried within the

FIG. 7.—GATE OF LIONS
AT MYCENÆ.

precincts of the Acropolis, and each man defended
himself as best he could.

Græco-Pelasgic architecture, however, is susceptible
of a more careful finish than the rude construction
of these walls might indicate.   Monuments of the

Achæan period, anterior to the Dorian invasions,
testify to a certain degree of art, and reveal that
mixture of individual and Oriental styles which
has already been remarked. The most beautiful
specimen of this architectural decoration has long
been known ; it is the sculpture which decorates the
tympanum of the Gate of Lions at Mycenæ. A bas-
relief represents two lions facing each other, on
opposite sides of a column with a circular capital, and
with a base—a design essentially Asiatic in origin.
The heads of the lions, undoubtedly of bronze, have
disappeared. Architectural fragments discovered at
Mycenæ by Dr. Schliemann show a keen sense
for decoration in art : here are fragments of fluted
columns, of friezes, and of shafts of porphyry ornamented with spirals and palm leaves.

The most remarkable remains of this period are
the so-called Treasuries of Orchomenus and of Mycenæ, which were probably of the nature of tombs.
Before Dr. Schliemann's excavations, the Treasury of
Atreus alone was known at Mycenæ ; but recent examination has brought a second to light. These
structures are built of slabs placed horizontally, the
layers of which gradually approach each other and
form a sort of pointed arch with a keystone. The
door, with splaying jambs, has a pyramidal form, and
is surmounted by a triangular tympanum. The interior
wall, faced with bronze plates, in the Oriental style, was,
without doubt, ornamented with columns ; near the
Treasury of Atreus has been found a column with circular base, and ornamented with chevrons and spirals.

Such were the obscure beginnings of an art which was destined to a brilliant development after the period of warfare had passed, and after the Hellenic people, firmly established in their own territory, were no longer obliged to provide solely for defence in their principal public works.

CHAPTER II.

THE ORDERS OF GREEK ARCHITECTURE — THEIR ORIGIN AND PRINCIPLES.

KUGLER : *Geschichte der Baukunst*, 1854—1873.
LÜBKE : *Geschichte der Architektur* (1855), 1875.
J. FERGUSSON : *The Illustrated Handbook of Architecture*, 1855, Second Edition, 1859. *History of Architecture*, Second Edition, 1874.
ERN. WAGNER and G. KACHEL ; *Die Grundformen der antiken classischen Baukunst*, 1869.
E. VINET : *Esquisse d'une Histoire de l'Architecture classique*, 1875.
CH. BLANC : *Grammaire des Arts du Dessin*, Third Edition, 1876.
CHIPIEZ : *Histoire critique des Origines et de la Formation des Ordres grecs*, 1876
BEULÉ : *Histoire de l'Art grec avant Périclès*, 1868.
HAUSER : *Styllehre der architect. Formen des Alterthums*, 1882.
BÖTTICHER : *Die Tektonik der Hellenen*, Second Edition, 1873—1881.
PEARSE : *Principles of Athenian Architecture*, 1851. (New Edition in preparation.)

---

§ I. ORIGIN OF THE ORDERS.

SOON after the "Dorian invasion" of Peloponnesus, Greek genius began to create those forms in architecture that are peculiar to it, and that bear its distinct imprint. It was by this step, by the creation and use of the "orders," that it discovered, as it were, and asserted its individuality : at the same time the principle of proportions, necessarily involved, gave to Greek architecture a beauty altogether unique and original.

The orders of architecture were finally developed at the close of the seventh century B.C., and during

the sixth century B.C. This was only after a long
period of experimentation, during which the several
elements of Hellenic architecture, borrowed from the
East, were applied somewhat at random, although at
last made subject to fixed laws. Before this epoch
ancient Hellenic edifices, erected either according to
traditions from abroad, or under purely local influences, may be classed in five distinct groups,
the characteristic types of which were simultaneously
followed throughout this early period. M. Chipiez has
defined these groups as follows :—

(1) Temples made of metal, or faced with metal—
in Media, in Judæa, and in Asia Minor. Greek
authors, especially Pausanias, speak of edifices constructed of bronze, such as the legendary temple of
Apollo at Delphi, that of Athena Chalciœcus at
Sparta, and the treasury of Myron, tyrant of Sicyon.
In the *Æneid* of Virgil the temple erected at Carthage
by the Phœnician Dido is described as being of bronze.

(2) The temple of wood, hardly more than an
enlargement of the cabins built of logs and clay in
which the ancient Hellenes dwelt. In later times
popular piety preserved these buildings with care.
Such were the temple of wood at Metapontum, and
the *sekos* of Poseidon Hippios, near Mantineia, which
tradition ascribed to the legendary architects Aga-
medes and Trophonius. Hadrian caused the latter
building to be enclosed within a marble temple. It
is probable that the use of wood was determined by
the poverty of the primitive cities.

(3) Temples in which both wood and metal were

used, wood in the upper part of the edifices. Instances of this type are the temples of Zeus at Nemea, and of Zeus Larissæus at Corinth. The wood became rotten in the course of time, and the roof was thus destroyed. Pausanias mentions several sanctuaries thus deprived of covering.

(4) Temples in the form of a cave : an instance is the sanctuary of the Delian Apollo on Mount Cynthus, in the island of Delos.*

(5) The temple of stone, forming a quadrangular enclosure, like that of Mount Ocha in Eubœa.

It is doubtful by what process the Greek temple, with its regular observance of the orders, came to be substituted for these irregularly constructed sanctuaries. Several explanations have been proposed. The oldest is that of Vitruvius, which prevailed during the whole of the Renaissance, and has been accepted more or less absolutely by the writers of our own times. According to this system, Greek architecture takes its origin from constructions in wood. Recent writers who accept this view (Hittorf, Beulé, Charles Blanc), while making due allowance for clear Oriental influences, have endeavoured to find the explanation of the several architectural members of a Greek temple in wooden constructions. They recognise in the entablature, or upper portion of the temple, beams, pegs, and ceilings of wood ; the column is derived from a wooden support squared and fluted by the axe. Another system, advocated by Viollet-

---

* Lebègue : *Recherches sur Délos,* 1876.

le-Duc and Regnault, sees the origin of Greek architecture in the necessary conditions of stone construction, and would thus regard the art as one born
on the soil of Greece. Discoveries made in the East
invalidate this theory, by showing that the elements
of the Greek orders were borrowed from Oriental
countries—Assyria, Phœnicia, and Asia Minor. M.
Chipiez has endeavoured to show that the principles
of architecture in wood but imperfectly explain
certain details in the Greek temple. According to
him, the forms adopted in the orders had been previously applied in Oriental buildings; in the final
determination of the orders, however, the Greek
artists were subject to the restrictions of plastic art,
and to the intuitive necessity of constructing a
reasoned and harmonious unity.

§ 2.   THE DORIC ORDER.

''The only orders purely Greek are the Doric,
Ionic, and Corinthian.'' The most ancient of these
is the Doric, which, towards the close of the seventh
century B.C., appeared simultaneously in all the
Dorian countries—at Corinth, Metapontum, Pæstum,
Segeste, Agrigentum, Syracuse. It is the national
order of the Dorian stock, which gave to it its own
characteristics of severity, force, and power.

In its essential elements, and at the time of its
perfect development, the Doric order was composed
of the following members :—The shaft of the column
rested directly on the stylobate or sub-base. It was

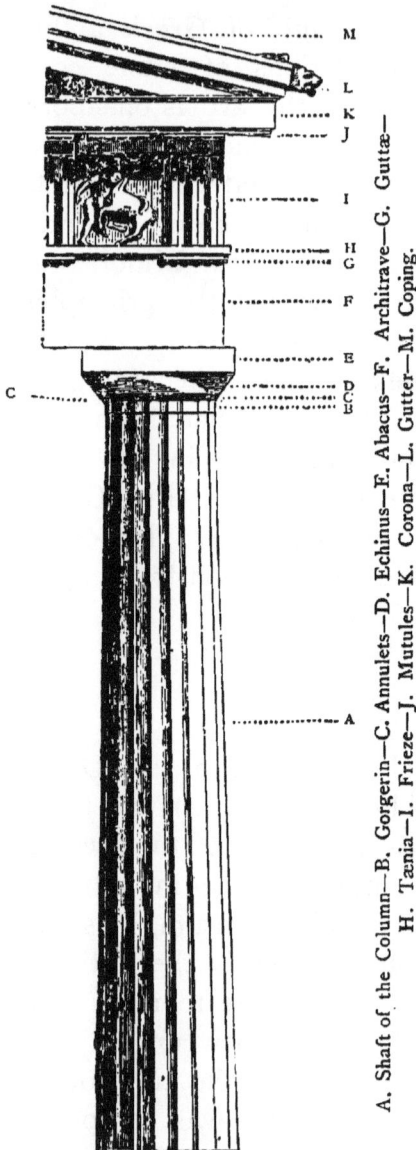

FIG. 8.—DORIC ORDER.
(Column with entablature.)

A. Shaft of the Column—B. Gorgerin—C. Annulets—D. Echinus—E. Abacus—F. Architrave—G. Guttæ—H. Tænia—I. Frieze—J. Mutules—K. Corona—L. Gutter—M. Coping.

cut with twenty vertical flutings without intervening spaces, and assumed to the eye the appearance of two truncated cones placed the one above the other at their largest section. The column showed in this way a swelling or *entasis*, which gave it the aspect of strength and of elasticity. It was composed of cylindrical drums attached to each other, which were often fluted after being placed in position. The top of the column terminated in the gorgerin or neck, enclosed between rows of fillets (in this place called annulets — "little rings"), which appear to unite as it were into a band the resisting forces of the shaft in order to support the capital. This is surmounted by a sort of cushion or *echinus*, upon which rests a flat rectangular block—the abacus—the edges of which project over the *echinus*.

This powerful column, with its capital, supported

FIG. 9.—DETAIL OF THE FRIEZE AND CORONA.

an entablature comprising several distinct members. First came the architrave, entirely uniform, composed of smooth blocks, with a free bearing from column to column.   Above this ran a flat moulding, the *tænia*, separating the architrave from the frieze.   The frieze was made up of alternate triglyphs and metopes. The triglyphs were bevelled flutings in decoration of a projecting rectangle resting directly on the tænia, beneath which were fixed six small marble cones called *guttæ*, or drops.

The channels of the triglyphs are in fact not three in number as the name seems to indicate ; they are made up of two complete ones on the face of the rectangle, and a half one at each edge.   These grooves have been explained in various ways.   Vitruvius, true to his principle, asserts that they are derived from the wooden tringles that were placed for decoration at the ends of the beams.   The metope is a slab of marble, sometimes smooth, but more commonly ornamented by sculptures in bas-relief, and it completely fills the space between the triglyphs.   In early times this space was left open, as we may infer from the statements of ancient authors.   In the Orestes of Euripides, the hero relates that he made his escape from a temple through the openings between the triglyphs.*

The entablature is crowned by the cornice, the essential feature of which is the corona or drip-stone, a surface made smooth so as to render easy the escape of water from the roof.   The lower, nearly horizontal, face of the corona consists of mutules, a sort of corbel, which support this projecting member.

---

* Euripides, *Orest.* 1369 *ff.*

They are covered with three rows of truncated cones, six in each row, termed *guttæ*. Above the corona runs the cyma, a waving beak-moulding, which on the sides of the temple terminates the entablature, while on the ends of the temple it forms the border of the pediment. The pediment is the large triangular member at the ends of the temple above the entablature, and is enclosed between two copings.

The Greeks termed the Doric the masculine order; in it nothing was sacrificed to mere grace. Its proportions are vigorous; its ornamentation soberly distributed; the general effect is one of power and austere simplicity, which might well suggest to the Greeks the robust outlines of the masculine form.

FIG. 10.—DETAIL OF THE MUTULES.

It is a commonplace among archæologists of the present time that the elements of the Doric order are to be found in Oriental architecture. At Karnak and in the columns of the Egyptian tombs at Beni-Hassan

we recognise the prototype, as it were, of the Doric
column ; the capital, composed of abacus and echinus
is found in Cyprus, at Golgos, and at Edde ; finally
the monuments of Pteria in Asia Minor show us
small structures surmounted by a pediment enclosed
with curved lines resembling outspread wings, which
seems to explain the name *aëtos* ("eagle ") given
by the Greeks to the pediment. But while we
admit this transmission of forms, made above all
through the mediation of Asia Minor and Phœnicia,
we must allow that Greek art has most remarkably
assimilated these elements, and has transformed them
to a degree that amounts to original creation. Monuments still in existence permit us to follow the steps
by which Greek genius approached perfection in this
respect.

The progress of the Doric order may be traced in
the study of dimensions, which were slowly modified
into what became canonical proportions. At first
massive and thickset, the columns had the effect of
heaviness, which was removed as the column became
more slender, the ratio of diameter to height gradually
increasing. The following table may give an idea of
this progress :—

UNCERTAIN DATE. Olympia, Heræum. The most ancient
Doric temple in Greece proper.
    Corinth. The column is not as much as even four
diameters in height : it is extremely heavy in aspect, and
the applied stucco increases this effect.

SEVENTH CENTURY B.C. (?) Selinus. The old temple ; height
of column, four and two-fifths diameters.

SIXTH CENTURY B.C. Selinus. The later temple ; height of column, four and one-half diameters. The Temple of Zeus, four and two-thirds diameters.

Syracuse. So-called temple of Artemis ; height of column four and two-fifths diameters. Temple of Athena (?) at Santa-Maria delle Colonne in the island of Ortygia ; height of column less than five diameters.

Pæstum. Great Temple of Poseidon, where the entasis of the columns is noticeable ; height of column, four and one-half diameters. Temple of Demeter, four and four-fifths diameters.

FIFTH CENTURY B.C. Ægina. Temple of Athena ; the height of the columns is five and one-third diameters. The age of perfection is near at hand.

Athens. The so-called Temple of Theseus ; five and one-half diameters. These are the proportions of the best period.

As the column grew more slender the entablature diminished in size, becoming less heavy and thus more in harmony with the column. The capital underwent similar changes ; at first somewhat flattened, sunken, and as if compressed under the abacus, the echinus gradually became more upright in its lines, and the curves became firmer. This progress can easily be measured by placing side by side for comparison two capitals, one from the old temple at Selinus, and the other from Ægina : the former of the period of beginnings, the latter near the period of perfection. (Figs. 11 and 12.) In fact, it was in the fifth century B.C. that Ictinus, in the construction of the Parthenon, and of the temple of Apollo Epicurius at Bassæ, employed the Doric in its most severe majesty. In the Propylæa, Mnesicles wedded the Ionic and Doric. In

E

course of time unsuccessful attempts were made to give grace to the Doric order, and these led to its decadence. In the fourth century it is still employed, but the Ionic tends to dethrone it. It was thus that Scopas, in the temple of Athena Alea at Tegea, chose the Ionic as the principal order, and relegated the Doric to the interior of the building. In Ionia a school of architects was formed which even proscribed the Doric order. Subsequently Roman architecture altered and perverted all the proportions that gave the order its original beauty : there is nothing in common between the feeble and heavy Doric described by Vitruvius and that of the Parthenon.

§ 3. THE IONIC ORDER.

According to ancient writers the Ionic order was of more recent origin than the Doric, and was used for the first time in Asia Minor in the temple of Ephesus, built by Chersiphron of Gnosus and his son Metagenes (Olymp. L., 580—577 B.C.). By these statements we are to understand that the temple at Ephesus marks the date when the canonical proportions of the order had become fixed, and this to such an extent that these two architects could write a treatise upon it. The Ionic order may, then, be said to have become fully developed by the middle of the sixth century B.C., and the testimony of ancient authors shows that it originated in Ionia.

The Ionic column differs essentially from the

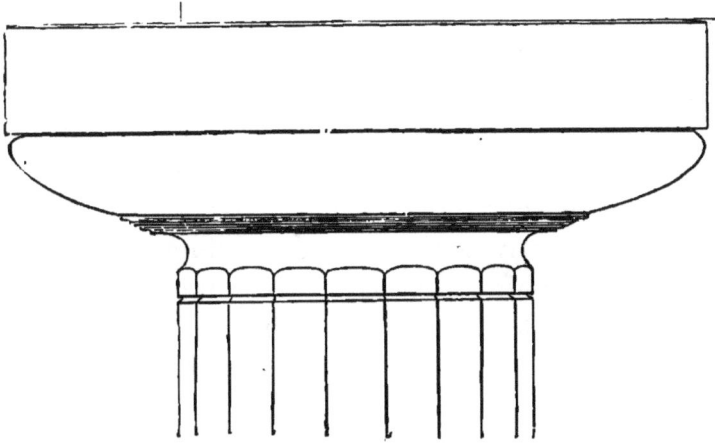

FIG. 11.—CAPITAL FROM SELINUS.

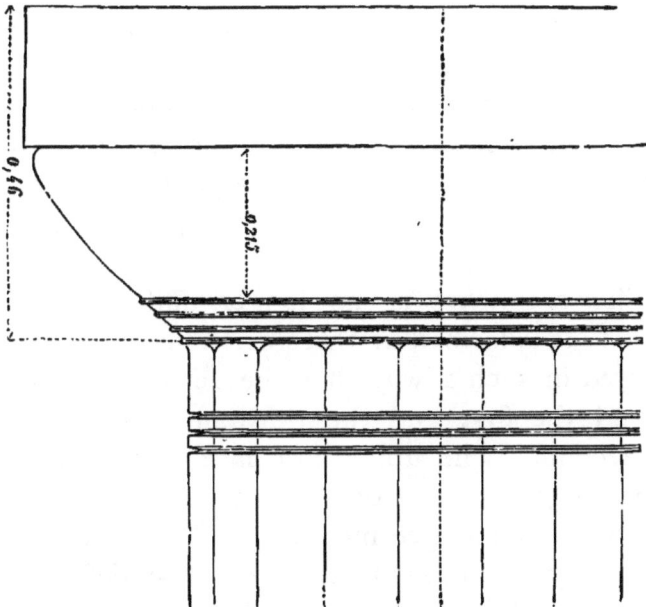

FIG. 12.—CAPITAL FROM THE TEMPLE OF ATHENA IN ÆGINA.

E 2

Doric column. Instead of resting directly upon the stylobate, it rests upon a base made up of a torus decorated with horizontal striæ, of a scotia or concave moulding, of a second torus, or sometimes a row of double rings. The base formed a sort of elastic cushion, the horizontal mouldings of which contrasted distinctly with the vertical lines of the flutings of the shaft. The latter were more deeply cut, but were of less width than in the Doric. They were separated from each other not by sharp edges (arrises), as in the Doric, but by narrow flat surfaces. The capital was fashioned according to a rectangular principle. The echinus was very diminutive, ornamented with ovules and rows of beads half concealed by the volutes, which spread widely on two sides : the thin abacus almost vanishes from sight between the volutes and the architrave. Occasionally a gorgerin or band, decorated with palm - ornament and aquatic plants, runs below the capital and forms the upper part of the shaft.

The architrave is no longer uniform as in the Doric. It is formed of three divisions or faces superimposed in such a way that the second juts slightly beyond the first, and the third juts slightly beyond the second. The uppermost face terminates on its higher edge in a row of beadings, and is united with the frieze by an ogee moulding. The frieze is ornamented by a continuous series of bas-reliefs, in imitation of Oriental buildings. Finally, the corona, entirely uniform, protects the entablature, and is terminated by a cyma decorated with beadings and with the egg-and-dart moulding.

This order, the rival of the Doric, was born in Oriental Greece, and became the national order of the Ionians, as the Doric had become the order of the Dorians in occidental Greece, where the Dorians were supreme. And yet we cannot maintain that the architects of the temple at Ephesus created it at one stroke ; in fact, its elements had long been in existence, and the Ionians on taking up their abode in Asia Minor had found them already in use. The Oriental origin of Ionic forms is no longer contested at the present day. Excavations at Nineveh and at

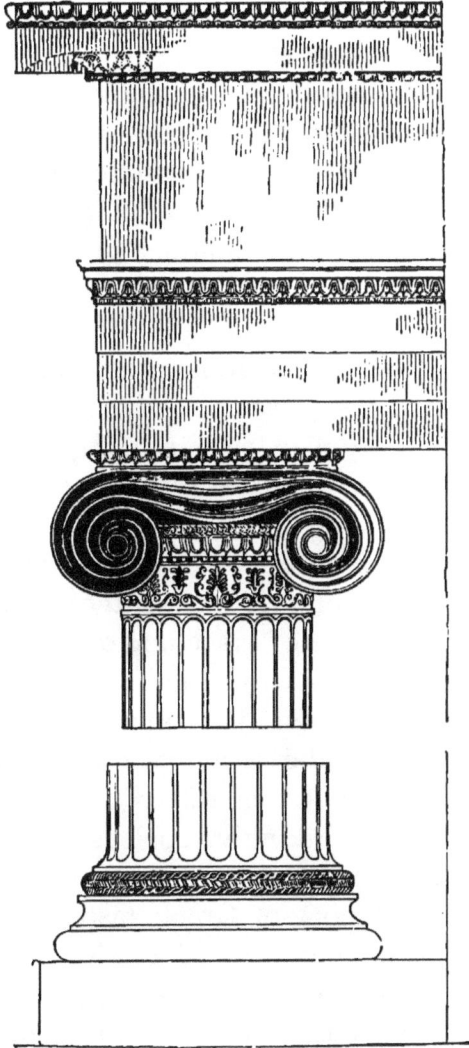

FIG. 13.—BASE, CAPITAL, AND ENTAB-LATURE OF THE IONIC ORDER.

Babylon, and discoveries made in Phœnicia, have
disclosed many monuments in which we may recognise a proto-Ionic type. It will suffice to mention
the bas-reliefs of the palace of Sargon at Khorsabad,
and those of Koyunjik, which give instances of
columns with volutes. In Phœnicia, and at Golgos,
we discover the principle of the Ionic capital, and a
bas-relief of Pteria in Asia Minor shows a small
building with columns surmounted by volutes, which
presents all the characteristic elements of the Greek
order.

Before the Ionic order had become independently
developed, the principle of the circular capital had
been applied with striking effect in the Doric; the
Ionic architects adopted the quadrangular capital
with volutes, and the column with base. In borrowing these forms from Oriental civilisations, however,
they impressed upon them their own native characteristics of grace and ornate elegance, which clearly
contrasted with the severe bareness of the Doric.
To carry on the comparison mentioned above, the
Ionic is the feminine order; its elegant and slender
forms lend themselves easily to rich ornamentation,
which strikingly contrasts with the austerity of
the Doric. The marble is finely carved wherever
the severity of Greek taste, ever opposed to excess,
would permit. Palm-leaves of exquisite form,
braids, lilies, and the egg-and-dart moulding, unite
marvellously with the curves of the volutes and with
the rounded forms of the base. Another essential
feature of the Ionic order is that, instead of being

immutable and inflexible, as is its rival order the
Doric, it is susceptible of an infinite variety. The
base admits of very diverse ornamentation. In Asia
Minor, in the temple of Apollo Didymæus, no two
bases are precisely alike; in one of them, for the
upper torus was substituted a cylindrical band decor-

FIG. 14.—DETAIL OF AN IONIC CAPITAL.

ated with palm-leaves; upon another the scotiæ gave
way to a dodecagonal trunk, each face of which was
decorated in a manner peculiar to itself. Occasionally, as at Ephesus, the shaft was ornamented with
sculptures. The capital finally shows great variations
in the combination of its lines, which may be grouped
under three heads :—

(1) The volutes of the capital are connected by a depressed curved line ; this is the classical style, used in the temple of Wingless Victory in Athens.

(2) The connecting line is an elevated curve, as at Phigalia.

(3) The connecting line is a straight line ; this was the form commonly used in Asia Minor.

This suppleness, in perfect keeping with the Greek instinct for freedom, has occasioned the just remark that the Ionic is the most Greek of the Greek orders. It is in fact the chief element among the innovations in architecture made by the architects of the age of Pericles. Passing in review the more important Ionic edifices, we shall find that this order was first employed in the temple of Artemis at Ephesus. This temple perished in a conflagration kindled by Herostratus, but remains of it have been discovered by Mr. Wood, among the foundations of the new temple erected upon the site of the old. The Heræum of Samos, begun about the thirty-fifth Olympiad (640 B.C.) by Rhœcus and his son Theodorus, was at first conformed to the Doric order, but was subsequently completed according to Ionic principles. It is possible that the Ionic was employed, in a partial degree, in the first Heræum, an Ionic column of the most ancient type having been found in Samos. In the fifth century B.C. it was almost exclusively employed at Athens. Ictinus had already shown its resources in the temple of Apollo Epicurius at Bassæ, near Phigalia. In Athens the order is most strikingly shown in the Propylæa, where it is united with the

Doric ; in the charming temple of Wingless Victory, an architectural gem with its limited dimensions ; finally in the Erechtheum, where it displays all its wealth and elegance.

The fourth century B.C. is, however, pre-eminently the Ionic century. This supremacy is clearly seen in

FIG. 15.—PALM-LEAVES, PEARL-BEADINGS, EGG-AND-DART AND LEAF-BUD ORNAMENT. (From an Ionic capital.)

the place given to the order in the external features of the temples. In Asia Minor the order blossomed forth in greatest perfection ; it attained its highest degree of excellence under the architect Pythius, who built the Mausoleum, and the temple ot Athena Polias at Priene. The masters of his school, Pæonius of Ephesus and Daphnis of Miletus, con-

structed the temple of Apollo at Didymi, where they
applied the Ionic with a marvellously rich inventive
power.   The flexibility of the principles of the order
did not protect it against radical modifications, introduced by the Asiatic architects, Hermogenes of
Ephesus and Thargelius of Tralles.   The former, in
the temples of Teos and Magnesia, removed one ot
the colonnades, and altered the profile of the bases ;
the latter, in the temple of Asclepius at Tralles, substituted the Corinthian for the Ionic capital.   Here
began the decadence of the Ionic order, which was
continued under the Roman architects.   The principles of this order, as formulated by Vitruvius, show
a singular degeneration.

§ 4. THE CORINTHIAN ORDER.

The youngest of the three Greek orders of architecture is the Corinthian.   In its canonical form it is
composed of a calathus, a sort of basket, about which
are applied the tall leaves of the acanthus, and of
helices, or volutes supporting an abacus less thick than
that of the Doric order, and concave on all of its four
vertical surfaces.   The projecting corners thus formed
require a support, which is given by the volutes.   The
entablature differs but slightly from that of the Ionic
order.

It is hardly necessary to recall the legend by which
the Greeks sought to explain the origin of this order.
A young Corinthian girl died, and her nurse placed
upon the grave a basket, containing some articles held

dear in life-time, and covered it with a tile. In the following spring, the basket, surrounded by the leaves of an acanthus, which had grown there, inspired the sculptor Callimachus with the idea of the Corinthian capital.

Of this pretty story it is sufficient only to retain the name of the artist. Callimachus lived about the eighty - fifth Olympiad (440—437 B.C.). Before this time, the principle of the bell-shape, upon which this system rests, had been in use; the French expedition to the Morea discovered at Corone a very ancient type, in which a Doric abacus rests upon a calathus, decorated at its base with slender and pointed acanthus leaves. The modification of Callimachus, important

FIG. 16.—CORINTHIAN CAPITAL.

enough to be termed an invention, succeeded in fixing the canonical forms of the capital. This artist was principally a worker in metal; he made the golden lamp of the temple of Athena Polias in Athens. We are thus justified in believing that the Corinthian capital conceived by him was of metal. The deep concavity of the acanthus leaves, the ornaments fastening the leaves to the calathus in such a way as

to hide the heads of the nails—in fine, all the
workmanship of the capital—seems to confirm
this theory. Furthermore, the capital of Callimachus was designed as the decoration of isolated
columns; the Greeks, struck by its beauty, made
an order of it.

About the ninety-sixth Olympiad (396 B.C.), in the
temple of Athena Alea at Tegea, Scopas employed the
Corinthian order for a portion of the columns in the
interior. Already, in 431 B.C., Ictinus had used the
order in the temple at Bassæ, for a column without
doubt intended for an inner sanctuary. These were,
however, but isolated experiments. The first instance
where the Corinthian order was openly applied for
the exterior seems to have been the small choragic
monument of Lysicrates in Athens, the date of which,
as indicated by its inscription, is the second year of
the one hundred and eleventh Olympiad, or 335 B.C.
About the same time, it is seen in the Didymæum of
Miletus, where it crowns the inner columns of the
façade. Thargelius, finally, was the first who, in the
temple of Asclepius at Tralles, built a colonnade of
Corinthian columns around the temple, and thus consecrated the use of this, the youngest of the Greek
orders.

The Corinthian order flourished vigorously in the
Roman epoch. It is beyond our task to trace this
later history, but it is nevertheless interesting to note
the fact, that in Roman edifices are found fine
details in execution that confirm the theory of its
origin in metallic forms. In Rome, for example, the

order followed in the interior of the Pantheon of Agrippa was Corinthian, where the columns were surmounted by metallic capitals ; and the portico of Cn. Octavius, erected in 147 B.C., after a victory over Perseus, was, says Pliny, "called Corinthian because the capitals of the columns were of brass."

CHAPTER III.

THE MONUMENTS OF ARCHITECTURE.

§ I. THE TEMPLE.

IT is above all in the religious architecture of the Greeks that we can study the use of the orders, the history of which we have sketched in the preceding chapter. The Greek temple is an organic whole—the highest expression of Greek art : painters and sculptors united in beautifying the dwelling of the god, in making it a harmonious whole, the unity of which was fixed by rules of the utmost precision.

The choice of the site for a temple was not left to chance. Usually, according to legend, the divinity indicated by some visible sign the place where he wished his sanctuary to be erected. Thus is explained a not infrequent occurrence—the existence of temples far from any human habitations, as at Ægina, Sunium, and Didymi. The building, facing eastward, was surrounded by a sacred precinct, the *temenos*, within which the piety of the faithful accumulated votive offerings, *stelæ*, and statues. On passing through this enclosure we find ourselves in the direct presence of the temple, the several features of which demand detailed description.

The proportions.—The fundamental element of the Greek temple is the sanctuary proper (*naos*, or *cella*), surrounded by architectural decorations which were variable within certain limits prescribed by the principle of proportions. We have already indicated what influence the colonnade, or *pteroma*, of the wooden hut was able to exercise upon Greek architecture. Greek temples may be classified according to the application, more or less complete, of the principle of colonnades; they may also be grouped according to the ordering of their columns.

The temple is said to be *in antis* when the chief façade is decorated by two columns, and when the two extremities of the façade are formed by the prolongation of the side walls of the *cella*, terminating in pilasters. It is *prostyle* when for these pilasters are substituted columns independent of the walls of the *cella; amphiprostyle* when it has a façade at the rear similar to that in front.* In *peripteral* temples the colonnade is prolonged along the lateral walls, thus passing completely around the cella; the *dipteral* temple shows a double colonnade around the cella ; the *monopteral* temple is round and has a circular colonnade, which supports a cupola; it is without interior walls or cella ; this form is rare in Greece. These simple principles were modified by certain architects, notably by Hermogenes, contemporary with Alexander, who adopted for the temple of Artemis at Magnesia the *pseudo-dipteral* principle, in which a second row of columns was attached to the walls of the cella.

* See below the plan of the Parthenon, p. 69.

Another classification is based upon the number of columns in the façade. / With four columns in the façade, the temple was called *tetrastyle*; with six, *hexastyle*; with eight, *octostyle*. The *decastyle* temple had ten columns, and the *dodecastyle* twelve. /

The originality of the Greek temple consisted above all in its scale of proportions; it was by this that the Greeks impressed upon their edifices a personal and individual character, with a scientific skill which the discoveries of archæology only confirm and attest. When Egypt was opened to the Greeks, in the middle of the seventh century B.C., the sensitive and keen spirit of the Greeks was struck with the aspect of power and force given to the Egyptian temples by their thickset columns, placed close together. But while the Egyptian temple had only dimensions, that of the Greeks had proportions, based upon the ratio of the several parts of the building to each other, expressed in terms of the diameter of the column taken at the base.

As applied in determining the width of the intercolumniations, this principle suggested a new classification of Greek temples. / The *pycnostyle* temple had intercolumniations of the width of one and a half measures, *i.e.*, in the *pycnostyle* between two columns, the diameter of the column at its base may be placed one and a half times; in the *systyle*, twice; in the *eustyle*, two and a quarter; in the *diastyle*, three; in the *aræostyle*, more than three times.

Variations in the height of the columns and of the entablature are controlled largely by these differences:

all the parts of the temple are subject to the same principle to such an extent that it is often quite possible completely to restore an ancient temple, with almost perfect accuracy, from the débris rescued from time.

The temple exterior.—The general features of the exterior of the temple were not uniform. For the sake of greater precision it will be well to study them in a selected instance. For that purpose we have chosen the Parthenon as a type.*

This temple rests upon a base reached by three high marble steps, which directly supports the shafts of the Doric columns ; these surround the main edifice, which has the form of a large rectangle. The architectural ornamentation is of the severe type of the Doric order, triglyphs, guttæ, mutules. Probably about the time of the orator Lycurgus a row of golden shields was hung upon the architrave of the eastern façade. The sculptured metopes alternating with the triglyphs, like so many square pictures, depicted a series of subjects borrowed from ancient legend, dear to the Athenians—the combats of the Lapithæ and Centaurs, the myth of Erechtheus and Pandrosus, the story of the fabulous beginnings of Athens, and the legend of Athena. The pediments were ornamented with statues in high relief, the work of Pheidias and of Alcamenes, representing, on the east, the birth of Athena ; and on the west, Poseidon and Athena disputing for the possession of Attica.

* Compare Michaëlis, *Der Parthenon*, with atlas, 1870--71.

F

Under the colonnade, on the upper part of the wall of the cella, a continuous frieze was carved; this represented the Panathenaic procession, with the priestesses of the goddess, maidens (*arrhephori*), the group of victims destined for sacrifice, armed war-chariots, and a long line of galloping horsemen, whose chlamydes float in the wind.

The plastic decoration of the exterior is completed by the ornaments placed in the highest parts. These were the gutters terminating in gargoyles of the shape of lions' heads, intended to empty the water above and beyond the corona. The extremities and the apex of the pediment were ornamented by *acroteria*, a sort of pedestal supporting various figures, sphinxes, vases, tripods, victories, lions. This arrangement, while not universal, was frequent; it is to be found in the Parthenon, in the temple of Wingless Victory, at Ægina, etc. In the temple of Ægina the ornament decorating the apex of the pediment has been recovered, as well

FIG. 17.—ACROTERIUM AND GUTTER.

as two figures of draped women that flanked this ornament. Finally we know that brilliant colours, judiciously disposed, distinctly accented the architectural details. This subject, however, will subsequently receive attention when we treat of *polychromy*.

In order to form an adequate conception of the external aspect of an ancient temple, we must see

FIG. 18.—PEDIMENT OF THE TEMPLE OF ÆGINA.

it in its surroundings, under a brilliant sun, which, by its distinct lights and shades, brings out clearly the finest edges cut in the marble ; we must encompass it with a girdle of mountains, the contours of which harmonise with the horizontal lines of the edifice, or contrast with the vertical columns. Placed on a naked rock, levelled only about the temple, the edifice appeared as a perfect work, complete in itself. The Greek had no trace of the peculiarly modern desire to give effect to a monument by the

F 2

symmetrical subordination of neighbouring monuments.   There is nothing more irregular than the
arrangement of the buildings of the Acropolis of
Athens.   The Propylæa which formed the approach
to the Parthenon are not in its axis.   The temple is
thus wholly independent, and owes its great beauty
entirely to the harmony of its own parts.

Modern research has shown how far this reasoned
endeavour after artistic perfection was carried.   Everything was arranged that no violence should be done
to the most delicate sensibility in vision.   From the
studies of Pennethorne, Penrose, and Paccard, we
learn that the Parthenon was, in its main outlines, a
sort of truncated pyramid.   In order that the vertical
lines of the temple might appear perpendicular,
Ictinus, the architect, aimed to correct the natural
errors of vision ; he inclined the walls of the cella and
the axes of the columns inward ; the columns at the
corners were made somewhat thicker and heavier, and
thus, surrounded as they were with light and air, they
no longer had the appearance of thinness in comparison with the other columns.   On the other hand,
the antæ, the cornices, the faces of the coronæ,
instead of sloping towards the interior, inclined out-
ward, so as to present to the eye of the spectator the
painted ornamentation with which they were covered.
The horizontal lines of the edifice had a slight curva-
ture ; the lines of the platform, those of the architrave,
and those of the corona which ran beneath the pediment, were lower at their extremities, thus forming
convex arcs.   The Greek architect took lessons from

the natural curves in the landscape, the curves of the mountains, and of the sea.

The temple interior.—The interior of the temple had three main divisions: the *pronaos*, the *naos* or *cella*, and the *opisthodomos.* On passing between the first row of columns of the peristyle directly under the pediment, we enter the pronaos; this division is formed by the prolongation of the lateral walls of the cella and a transverse wall. (Fig. 19, C.) A row of columns formed the frontage of the pronaos; a lattice grating between the columns closed it when necessary, and made secure the objects placed within the enclosure. The naos, or cella (D), to which the pronaos gave access, was pre-eminently the abode of the god. It was divided into a nave and two aisles by two rows of superimposed columns. In each of the rows the lower tier of columns, of the Doric order, rested

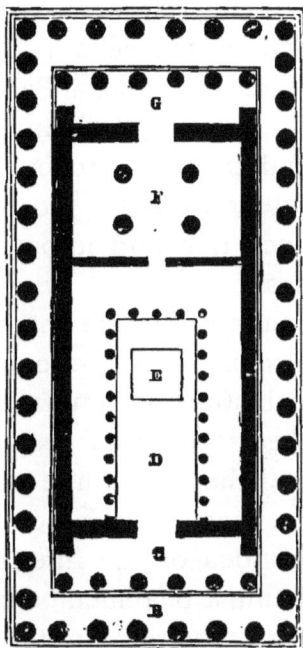

FIG. 19.—PLAN OF A GREEK TEMPLE. (Parthenon.)

directly upon the pavement, as may be proved from the traces left on the flagstone of the fluting of the columns, which was done when they were in position; this lower tier supported an architrave upon which

rested another tier, which was sometimes Ionic, sometimes Doric.   It is not known whether or not this
second story was floored, thus forming a sort of gallery.
In the Parthenon Paccard has failed to discover any
traces of stairs, which such a gallery would require
as a means of access.

The statue of the divinity stood at the further
end of the cella (E).   In the Parthenon it was the
statue of Athena Parthenos, one of the masterpieces
of Pheidias, brilliant with costly metals, with ivory
and precious stones; it rested upon a pedestal delicately carved.   To form an idea of the aspect, at
once rich and imposing, of the sanctuary of the goddess, we must bring before the imagination the
columns of the naos decorated with shields and
armour; the store of works of art, votive tablets,
rich stuffs, accumulated about the pedestal of Athena;
the statue itself resplendent with the sheen of gold
and the dead white of ivory.   Byzantine churches—
St. Mark's at Venice, for instance—can alone give us
an idea of the interior decoration of a Greek naos.
It must be remembered that this ornamentation had an
infinite variety in its details.   At Olympia the statue
of Zeus, seated upon a throne of gold, ebony, ivory,
and marble, was enclosed within low barriers covered
with paintings; the floor was set in black marble
destined to receive the oil poured upon the ivory of
the statue to preserve it.   At Didymi the statue of
Apollo stood under a small shrine.   In Athens, in
the Erechtheum, a golden lamp, the work of Callimachus, was placed before the wooden statue of Athena

FIG. 20.—COLUMNS IN THE INTERIOR OF THE NAOS.

Polias ; this lamp was shaded by a bronze palm-tree, which served to carry the smoke to the roof.

The third and last main division of the temple was the opisthodomos. (Fig. 19, F.) Though sometimes completely isolated from the cella, the opisthodomos usually communicated with the naos by means of an entrance lying in line with the entrances from the pronaos ; this space was filled with a bronze door in the midst of a grating. In the Parthenon it is probable that the ceiling of this part of the temple was supported by four columns. In the opistho-domos were preserved the treasures of the goddess, made up of gifts, thank-offerings, the products of sacred property of all sorts, etc. ; here also were kept objects of historic interest, as the sword of Mar-donius, the throne of Xerxes with feet of silver, the treasure of the State, with the grand seals. These treasures were guarded by the "treasurers of the sacred riches of Athena," who made an inventory, every four years, of the property in their keeping. Marble *stelæ* have preserved to us the details of such inventories, not only of the Parthenon, but also of the Asclepieion of Athens, and of the temple of Apollo at Delos.*

---

* Inventories from the Parthenon: *Corpus Inscriptionum Attic-arum*, Vol. I. ; *Traditiones quæstorum Minervæ*. Inventories of the Asclepieion ; Girard and Martha, *Bull. de Corr. hellénique*, 1878. Accounts of the guardians of the temple of Delos : Homolle, *Bull. de Corr. hellénique*, 1878, pp. 570—584. The remarkable discoveries of M. Homolle, the results of which have not yet been published in full, make it possible for us to understand in its detail the administra-tion of the temple of Delian Apollo.

These inventories of the Parthenon mention golden crowns, couches from Chios, breast-plates, swords, lyres, ivory tablets, gilded shields, vases, statuettes of precious metals, etc. In cases of great financial stress the State occasionally made use of the private treasure of the goddess. Athena thus lent to Athens.

The general plan of the interior of Greek temples varied but slightly ; that of the Parthenon may be regarded as the prevalent type. Where, however, the temple was that of an oracle (*manteion*), as at Didymi, the arrangement differed. In this temple, sacred to Apollo, the pronaos was separated from the naos by an apartment where visitors who had come to consult the oracle waited until the god had inspired the Pythia, or priestess. This division of the temple was termed the *œcos* (οἶκος).

*The lighting of the hypæthral temples.**—Few questions have been more discussed than that of the lighting of Greek temples. How did the sunlight penetrate into those temples termed by the Greeks *hypæthral* (ὑπὸ αἴθερος)? This name gives us to understand that certain parts of the edifice were without outer covering, and exposed to the free air ; but how are we to reconcile this arrangement with the necessity of protecting from the weather the treasures accumulated in the sanctuary, and the image of the divinity itself? The ruins of the temples offer but little help in the solution of this problem ; as is well

---

* Beulé, *Histoire de l'Art grec avant Périclès*, p. 281 *ff.* ; Chipiez, *Mémoire sur le Temple hypèthre*, Revue archéol., new series, Vol. XXXV.; Fergusson, *The Parthenon*, 1884.

known, in the complete destruction of the upper portions of the temples no remains have been preserved of the roofing. The question, then, is one that can be solved only by the aid of theories based on the statements of ancient writers. Among these authorities the most important is Vitruvius, who gives the following description of the hypæthral temple :—" The hypæthral temple has ten columns in the pronaos and in the posticum. As in the great temples, the exterior colonnades of which are made of two rows of columns, the hypæthral temple has in the interior of the cella two rows of superimposed columns, running at a distance from the walls, and leaving a space for moving about, as in the porticoes of the peristyle. *The intermediate space is open to the sky, and is unroofed ;* at either extremity there are doors, both in the pronaos and in the posticum."*

Archæologists have put forth several theories aiming to harmonise the statements of the texts with the necessities of construction. According to one view, widely accepted, and adopted by Mr. Wood in his reconstruction of the Artemiseum at Ephesus,† the temple was lighted by an opening (ὀπαῖον) in the roof and ceiling. But this aperture, however small it might have been, would have permitted rain-water to fall into a portion of the cella. In that case we are obliged to suppose the use of an awning, or of translucent stones, which is a gratuitous hypothesis.

---

* Vitruvius, III. II. 8.
† J. T. Wood, *Discoveries at Ephesus*, 1877.

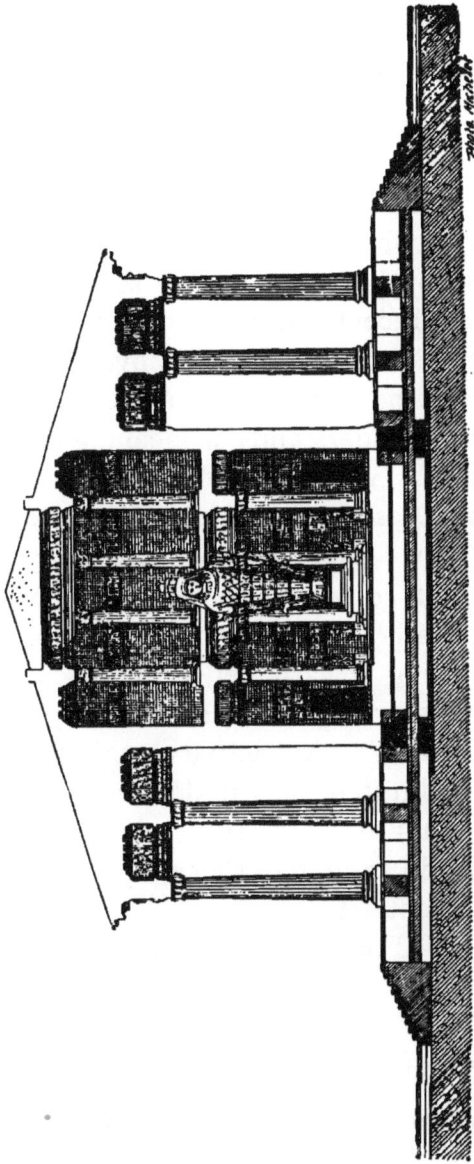

FIG. 21.—ARRANGEMENTS FOR LIGHTING THE TEMPLE AT EPHESUS.

(According to Mr. Wood.)

Fergusson and Canina* have supposed that a sort of
lantern, pierced with lateral apertures, might have
been placed above the opening. This arrangement
would have required an ungainly interruption in the
lines of the roof, while representations of temples on
medals show a continuous and unbroken line of roofing.† The most recent theory is that of M. Chipiez,
who has removed the difficulties by an ingenious
interpretation of the text of Vitruvius. "The intermediate space" he understands as meaning the interval
between the cella walls and the nearest series of
columns, forming a sort of aisle along the lateral walls
of the cella. "If we remove a row of the large marble
tiles in the roof [directly above this intermediate
space] the light falls first . . . upon the ceiling of these
aisles, then, passing inward between the columns of
the upper tier, is spread as from so many windows
throughout the nave." Attractive as this theory may
be, it must be admitted that it does not give a final
solution, and that the question demands further
examination, as do so many other problems raised by
the study of ancient art.

Polychromy.‡—For a long time the idea that

---

* Fergusson, *On the Temples at Ephesus* . . . . *as illustrating the
Hypæthreum of the Greeks,* 1877 ; Canina, *Architectura antica descritta,*
1844.

† Donaldson, *Architectura Numismatica,* 1859, 401.

‡ Hittorf, *Architecture polychrome chez les Grecs* (1830), 1851. Beulé,
Histoire de l'Art grec avant Périclès, p. 244. Valuable data for this question are furnished in the restorations made by the architects pensioned
by the French Academy in Rome. These restorations are preserved
in the École des Beaux-Arts in Paris.

painted decoration was applied to Greek temples was rejected as insulting to Greek art. The polychromy of the temples has been admitted first in our own day, and then only after long discussion. This is not the place to recall the several phases of the discussion, nor to show how modern prejudice has slowly yielded to the logic of facts. To Hittorf belongs the credit of having united the arguments in favour of polychromy, and of having clearly stated the question. His views, ardently adopted, and pushed to their extreme, have given rise to extravagant theories of polychromy, according to which every part of the temple was coated with striking colours, under which the whiteness of the marble entirely disappeared. A more moderate theory, representing a system of partial polychromy, admits the use of painting in a manner more restrained and more in keeping with the sober taste of the Greeks. Furthermore, it has been proved that polychromy varied according to the epoch, in a retrogressive movement: at first applied liberally to the stucco which coated archaic temples, it was reduced with the progress of time and with the better taste of the schools. Thus this subject of polychromy cannot be entered upon without taking into account periods, localities, and the various schools of architects.

The traces of painting* observed upon the various architectural members of the temples in Ægina, Athens, Sicily, and Magna Græcia, aid in the

---

* The painting was encaustic; the artists are styled ἐνκαυταί in an inscription relating to the Erechtheum. *Corp. Inscr. Attic.*, I. 324.

reconstruction of the painted decoration of the Doric temples of the sixth and fifth centuries B.C. In the time of Peisistratus the columns appear to have been painted a pale yellow; the colour was applied to the stucco coating of the stone, which offered a fine and smooth surface to receive it. It is not known whether it was customary to paint the capital; the capitals of the portico at Pæstum, however, should be mentioned, where the painted palm-leaves are still prominent, the remainder of the stone having been corroded by the sea-air. The architrave in Ægina was painted in a uniform red tint, which served as a background for the gilded shields, and for votive inscriptions in metallic letters. Above the architrave the frieze presented alternate triglyphs in blue and metopes in red; on the red background of the metopes, carved bas-reliefs, with their accessories of gilded bronze, stood out most distinctly. The mutules of the cornice were blue. In the pediment the tympanum had a blue background, throwing into relief the sculptured figures within it; the mouldings surrounding the tympanum were decorated either with red and green leaves or with red and blue leaves. Add to this, above the entablature, gutters coloured in vivid hues, tiles, acroteria, antefixes of marble or of terracotta, decorated with palm-leaves or with Gorgons' heads, and one may form some conception of archaic polychromy, with its decided tints, which are in perfect accord with the austere lines of the early Doric.

When the proportions of the temple became more elegant, and marble was substituted for stone, and in con-

sequence more finished work was required, the colours
were distributed less lavishly than before.   The Ionic
order especially employed polychromy with refinement
and moderation.   The delicate lines, the exquisite
chiselling of the marble,
which ran like lace
around the gorgerins of
the capitals and under
the voluted abacus,
could not be hidden
under a bed of colour.
The colour merely
outlined these features,
in order to bring them
out from the white-
ness of the marble,
flooded as it was
with glowing light ;
and to the vivid reds
and blues were added
the brilliancy of gild-
ing.   This is proved

FIG. 22.—ANTEFIX OF MARBLE.

by an inscription of the ninety-second Olympiad,
giving the account of some expenses of the Erechtheum.*   We find there mentioned " one hundred
and sixty-six leaves of gold at one drachma each,"
which were intended for the gilding of the eyes of the
volutes, and for the ornamental work in the ceiling.
But here, as in the Doric order, there was no rigid

* Corb. Inscr. Attic. I. 324.

system ; the use of polychromy varied with the tastes of architects and with the traditions of the schools.

Observations made upon the ruins of Ionic edifices at Priene, Didymi, Ephesus, Halicarnassus, and Athens, make it possible to note to what extent polychromy was associated with the Ionic order. Two colours, red and blue, were principally used. The former was reserved for backgrounds, and for parts in the shade, which it brought out with its deep tints : thus at Halicarnassus the ruins of the Mausoleum show us rows of carved pearls standing out from a red background ; sometimes red was employed to outline the egg-and-dart ornament,* on the darts, and on the shell of the ovules. Blue was applied to the higher, more lighted surfaces—to the bases of the ovules, for instance—while the more salient details remained white. From this arrangement a harmony of colours resulted, subdued and yet brilliant, warm and intense shadows, blues softened by the sunlight, and finally the exquisite carvings, preserving on their more salient surfaces the brilliant whiteness of the marble in all its purity.

The laws of polychromy can be rigorously established only after a minute study of all the data, which seems not yet to have been made. It can at least be said at present that in all points it was in perfect harmony with Greek genius. This system, so contrary to our modern tastes and to our views as to the

---

* Compare Newton, *Halicarnassus, Cnidus, and Branchidæ,* plate xxix., various mouldings with their original colours; O. Rayet *Le Temple d'Apollon Didyméen,* in *Gazette des Beaux-Arts,* 1876.

FIG. 23.—GUTTER OF PAINTED TERRA-COTTA.
(Athens.)

G

divisions between the arts, teaches us once more to what an extent, in the soul of the Greek, the most diverse arts could be made to unite for a common object. We can, furthermore, but imperfectly understand Greek polychromy, if we fail to remember the conditions of climate which rendered it almost necessary. In that golden light, the uniform tint of marble would have been monotonous; details would have been lost in the unbroken white sheen, which the brilliancy of the summer sun would have rendered almost blinding. It was necessary to accent the chaste lines of the edifice, to give distinct effect to the details of exquisite workmanship, and to brighten them by brilliant hues, which wonderfully harmonised with the clear radiance of the sunlit heavens.

§ 2. THE PROPYLÆA. PORTICOES.

The idea of erecting in front of temples and other great edifices a structure in decoration of the approaches (*propylæa*) was not peculiar to the Greeks. The Egyptians and the Assyrians had already erected such buildings. Among the Greeks, propylæa were of frequent occurrence, as in Corinth, at Priene, in front of the temple of Athena at Sunium, and at Eleusis. The most famous were those on the Acropolis of Athens, which shared with the Parthenon the admiration of all Greece. "The Propylæa and the Parthenon," says Philostratus, "would have been glory enough for Pericles." Begun by Mnesicles in 437 B.C. (in the archonship of Euthymenes), and five years in constructing,* they represented the perfection of civil architecture in Athens. Their plan is very simple. The chief feature was a wall pierced by five doors, the central one of which was the greatest in size, admitting chariots and horsemen. In front of this wall stretched a hexastyle portico between two lateral

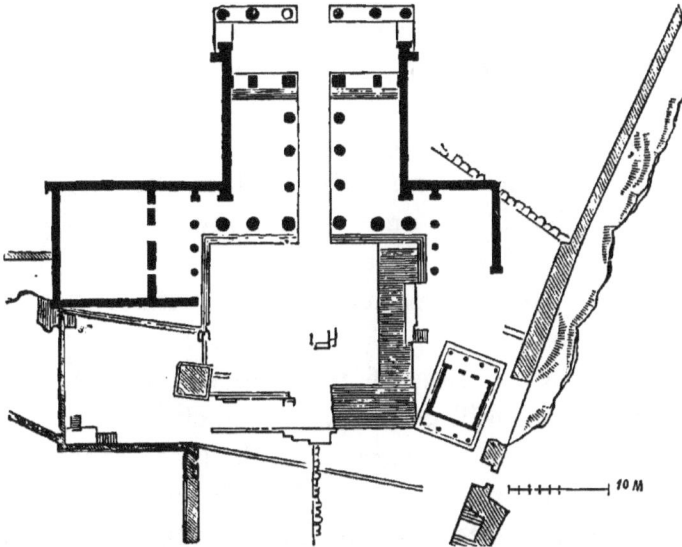

FIG. 24.--PLAN OF THE PROPYLÆA OF THE ACROPOLIS OF ATHENS.

wings, each of which had its portico, that on the right forming a hall called the *Pinacotheca*, still partially in existence. The façade of the wings presented merely a wall ornamented by a frieze of triglyphs. †

* The Propylæa were never finished ; one may to-day see unhewn portions on the surfaces of blocks of marble.

† Compare Beulé, *L'Acropole d'Athènes*, 1863, Bk. I., chapter vi.; Bohn, *Die Propyläen der Akropolis zu Athen*, 1883.

G 2

The rear wall, looking towards the Parthenon, was likewise provided with a portico, not so deep as the outer one, but with the same number of columns. This plan, simple as it was, was executed with singular originality. We have already remarked that the columns of the two lateral porticoes were of smaller dimensions, that they might not throw into the background those of the principal part of the building—the façade; the doors were arranged in a descending scale, those of larger size being in the centre, as required by the principle of harmony and unity. Finally, by a bold and successful stroke, Mnesicles employed the Doric and Ionic simultaneously, the former in its severe beauty, the latter in its chaste ornateness, each in perfect accord with the other. One element of beauty in the Propylæa which provoked the admiration of the Greeks, was the rare excellence and skill of the workmanship. Pausanias was astonished at the enormous blocks of marble put together with infinite pains: "The Propylæa," he says, "are made of white marble; they are the most admirable work executed up to the present time, both for the size of the blocks employed and for the beauty of workmanship."[*] As if to give greater prominence to this perfection in technique, Mnesicles provided the Propylæa with a sub-base of common stone from the Peiræus, thus despising accessories and searching for beauty in the work itself.

The object of the Propylæa has been a theme of

---

[*] Pausanias, I. 22.

discussion. Are we to see in them a work for defence, built to protect the approaches to the citadel of Athens, or rather a decorative structure? The latter, which justifies the perfection of the work, is probably the correct view.

Of the porticoes which surrounded the market-places, of the gymnasia and the palæstræ, there remain only masses of débris. In Athens the most important ruins of works of this class are of the Roman era. Such, for instance, is the structure which marks the entrance to the agora, or market-place, of Athens; of this there remain only four Doric columns supporting a pediment. An inscription on the architrave indicates that the edifice was built by the liberality of Julius Cæsar and of Augustus, in the archonship of Nicias, and was sacred to Athena Archegetis. Another inscription, mentioning an edict of Hadrian on the sale of oil, confirms the hypothesis that these ruins are of a portico in the agora.*

To the Roman era likewise belongs the *stoa* or portico of Hadrian, the façade of which, with Corinthian columns, is still in existence. The stoa was a quadrangular enclosure, surrounded by porticoes, with halls decorated with paintings and statues. It had, besides, a temple to all the gods, or Pantheon, and a library. Athens, like all Greek cities, possessed a large number of these edifices clustered in the neighbourhood of the agora, in the centre of public life

* There can be seen on the site of ancient Pompeiopolis, in Cilicia Trachea, a beautiful colonnade which ran along a street or *dromos*, undoubtedly near the agora of that city.

and business. We know, more from the descriptions
of ancient writers than by the ruins, of the portico of
Hermes, the gymnasium of Ptolemy, the portico of
Attalus, and the Poikile, or Painted Porch. Here it was
that Athenians passed the greater part of the day,
with their fondness for life in the open air, for movement, and for political discussion. Under these porticoes, ornamented with statues and Hermæ, they
discussed the questions of the day, or treated of
the affairs of the State. Nothing was neglected in
the embellishment of these popular edifices. The
Poikile was ornamented with paintings by Panœnus,
Polygnotus, and Micon, representing the principal
myths of Attica, or the exploits of the Athenians in
former times : Theseus fighting with the Athenians
against the Amazons ; the Athenians at Œnoë, ready
to engage in battle with the Lacedæmonians; finally
the battle of Marathon, where the protecting divinities
of Attica fought in the front rank of the Athenian
army.[*] Throughout his life, spent for the most part
out of doors, there never was a moment when the
Athenian had not before his eyes some memorial of
some glorious deed in the history of his country.

The life of the Greek was divided between public
affairs and religious ceremonies. The history of the
discipline of the Ephebi, which was but a preparation
for the duties of citizenship, shows how numerous
were out-of-door occupations.[†]

---

[*] Pausanias, I. 15.
[†] Compare A. Dumont, *Essai sur l'Éphébie attique*, 1876.

The Greek dwelling-house was accordingly of the greatest simplicity, and was constructed of the humblest materials. Ruins of houses, still visible at Stymphalus in Arcadia, at Syracuse, on Pnyx Hill in Athens, show how mean was the private residence of the Greek. The only floor was the level rock, which was also often utilised as a wall. On this foundation the house was raised ; the ground floor was reserved for the men of the family ; the first floor, or *gynœceum*, for the women ; the whole was of singularly contracted dimensions. A Greek city offered a curious contrast between the splendour of its public buildings and monuments, and the humble aspect of its private dwellings. It was only at an epoch comparatively late, and under the influence of Asiatic manners, that luxury was introduced in private buildings. Demosthenes complained that the houses of private citizens rivalled public edifices in their magnificence. "If you consider the houses of Themistocles, of Miltiades, and of great men of those times, you will see that they were in no way more magnificent than were the houses of the majority . . . But in our day so great is the wealth of many citizens engaging in public life, that numbers of them have built private houses of greater magnificence than are many even of our public buildings."* Still later, in the Alexandrine age, the refinements in the arrangements of a house were carried to their highest pitch. We may judge of this from the description given by Vitruvius of a

* Demosthenes (XXIII.), *Contra Aristocr.*, 207, 208.

typical Hellenic house, which contained porticoes with columns, bath-rooms, a picture gallery, exedræ, banquet halls, and library. The passage from Vitruvius makes us understand how the influence of Ionia had developed among the Greeks the taste for luxury and comfort; "to build a house after the Ionian manner" was a common saying.

§ 3. THE THEATRES.* THE ODEONS. THE STADIUM.

Ruins of theatres are frequently met with in Greece and in Asia Minor: remains are preserved, more or less complete, of the theatres of Tlos, of Patara, of Caunus, of Telmissus, of Cnidus in Asia Minor, and of Sunium, of Epidaurus, of Argos, of Sparta, etc., to mention only the more important. As a type we select the theatre of Dionysus in Athens; † it gives, with interesting peculiarities of its own, the arrangements commonly adopted in this class of constructions. This Dionysiac theatre has been well known only since the excavations begun in 1862 by Strack, a German architect. Situated at the south-eastern angle of the Acropolis, it lies in part in the bed of the rock, from which some of the tiers of stone seats were excavated. The southern wall of the Acropolis overlooks it. It is thus

---

* Wieseler, *Theatergebäude*, 1871; O. Benndorf, *Beiträge zur Kenntniss des attischen Theaters*, 1875; Ottfried Müller, *History of Greek Literature* (tr. by Lewis and Donaldson, 1850).

† Compare Wheeler, *The Theatre of Dionysus* (*Papers of the American School at Athens*, Vol. I.), 1885.

represented on an Athenian coin or medal, where the engraver has not failed to indicate the imposing view of this side of the Acropolis : the rocky crags pierced by a grotto, the wall running along the crown of the hill, and above that the majestic profile of the Parthenon. The interior plan of the theatre of Dionysus seems to have served as a model for other edifices of this kind. A brief description will thus give a clue to Greek theatres in general. The theatre comprised three main parts : (1) the *scene*, or stage, (2) the *orchestra*, and (3) the *theatre* proper, *i.e.*, the tiers of seats reserved for the spectators.

The stage had been repaired and remodelled several times, first in the reign of Hadrian, and afterwards under Septimius Severus, by an architect named Phædrus. The object of these modifications was to give greater width to the stage at the expense of the orchestra. We shall here consider only the classical Hellenic stage, the walls of which, still in good condition, are visible : this stage dates from the administration of Lycurgus (330 B.C.). It was at this time that the great orator, charged with the care of the finances of the Athenian republic, completed the work upon the theatre begun in 496 B.C. (Ol. LXX.) by architects whose names are unknown. This stage was situated far back, and left ample space for the orchestra. Like all Greek stages, it comprised two parts ; a rectangular space where the actors spoke their parts (λογεῖον), and the background (σκηνή), with a wall penetrated by three doors, the central one of which was called the *royal door*. At each of the two wings at the ends of the stage was another door, one of which was supposed to lead into the country, the other to the agora. Tragedies were represented in the midst of permanent decorations, remains of which have been recovered ; fragments of Doric columns, of architraves, of pediments, formed this decoration, which was thus of a simple and purely architectural character. Other painted decorations were placed in front of the rear wall. Complicated mechanical contrivances, which brought apparitions into view, or created other necessary scenic effects, were kept in a place below the stage (ὑποσκήνιον), divided into separate compartments, as may be seen in the theatre at Athens. The stage, a platform somewhat above the level of the orchestra (about one and a half metres), rested upon a wall decorated with sculptures. The wall of the stage, as reconstructed by Phædrus, has been recovered. Figures of kneeling satyrs support the *proscenium* after the manner of Caryatides: the intervals between these figures are filled up with bas-reliefs representing scenes from the life of Dionysus. A staircase in the centre connected the λογεῖον with the orchestra.

This part of the theatre, the orchestra, which separated the stage from the tiers of seats, had the form of a semicircle, and was paved with marble flags. In the centre rose the altar of Dionysus (θυμέλη), around which moved the chorus under the leadership of the coryphæus. The semicircle outlining the orchestra was defined by the front row of the seats, which, rising as in an amphitheatre, were divided

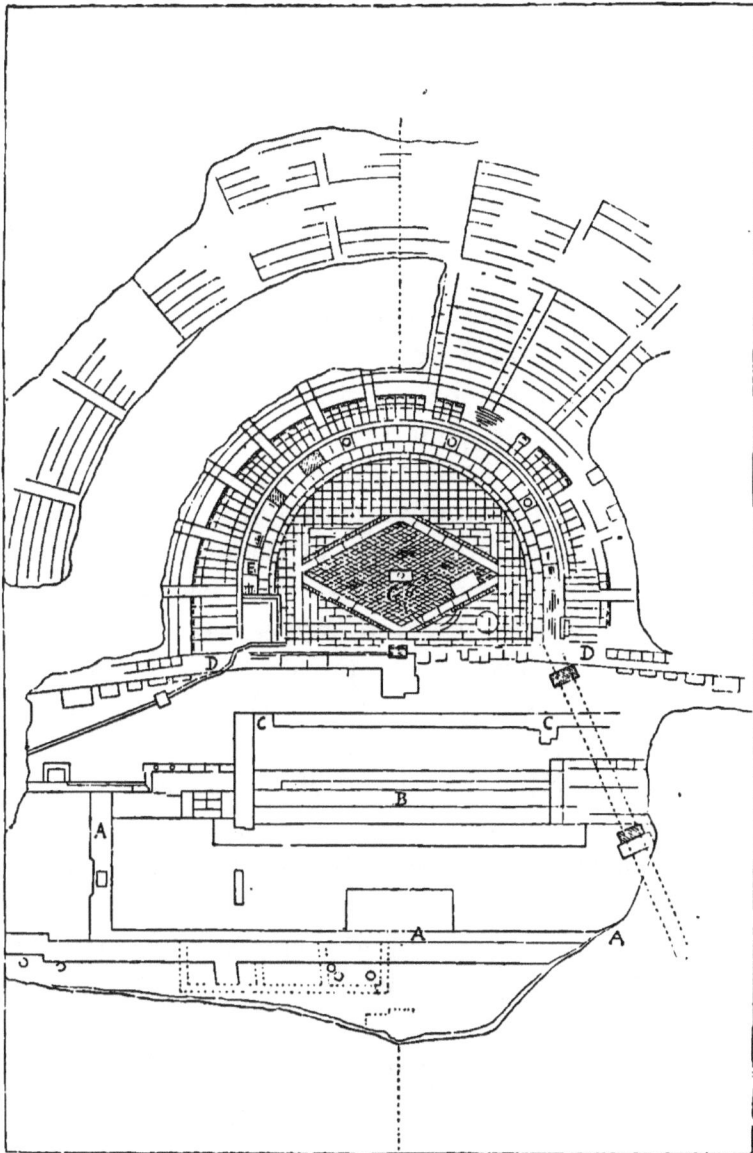

FIG. 25.—PLAN OF THE EXCAVATED THEATRE OF DIONYSUS.

A.—Hellenic stage ; B.—Stage of time of Hadrian ; C.—Stage of Septimius Severus ;
D.—Marble seats for those having the right of proëdria ; E.—Podium separating the
seats from the orchestra ; G.—Orchestra.

into compartments by flights of steps (κλίμακες). The tiers of seats were doubtless remodelled under Hadrian, for the number of compartments (κερκίδες) is thirteen, and thirteen was the number of the Attic tribes under this emperor. The front row of seats was reserved for the priest and for the archons, who had by their offices this right of προεδρία; the next two tiers were also occupied by people of note. These seats, having the form of easy arm-chairs, are of Pentelic marble; the rows are separated into seats by sculptured divisions; the seat itself is slightly hollowed to receive a cushion. An inscription, cut upon the lower front edge of the marble chair, in-dicates to what high personage the seat belongs, whether priestly or political.* Here appear the names—to mention only a few—of the priest of Olympian Zeus, of Zeus Polieus ("protector of the city"), of the Delian Apollo, of Poseidon Phytalmius, of the Pythian Apollo; also of the interpreter of the oracle of Apollo, of the hierophant who presided at the initiation into the Eleusinian mysteries, of the *hieromnemon*, deputy on the part of Athens to the Amphictyonic Council, and finally, of the archons (eponymous, basileus, polemarch, and the thesmothetæ). The chair which is directly in front of the stage is that of the priest of Dionysus Eleuthercus, the god to whom the theatre was sacred;† it was further distinguished from the

---

* *Corp. Inscr. Attic.*, Vol. III., 1. 240 *ff.* These inscriptions are a most valuable material for the study of the Athenian priesthood and of the hierarchy in general.

† It is inscribed ΙΕΡΕΩΣΔΙΟΝΥΣΟΥΕΛΕΤΘΕΡΕΩΣ.

others by a richer ornamentation. The back was
decorated by an elegant bas-relief representing satyrs

FIG. 26.—CHAIR OF THE PRIEST OF DIONYSUS ELEUTHEREUS.

in the archaic style; on the front of the seat was
sculptured a miniature frieze representing a combat
between two Arimaspes and griffins; finally on each
of the arms was the figure of Agon, the presiding

genius of cock-fights, which took place within the precincts of the theatre. The priest of Dionysus had thus the place of honour. We can easily understand the comical effect produced by one of the scenes in the *Frogs* of Aristophanes. When Dionysus, accompanied by his slave Xanthias, arrives in Hades, and is struck with terror at the monsters guarding the entrance, the actor in the rôle of the god turns towards the audience and exclaims, "O priest, rescue me, that I may become your comrade at the drinking bouts!* To the right and left of the stage opened large passage-ways, formerly decorated with statues; on the pedestals, which alone are preserved, we read the names of the most famous dramatic poets, Thespis, Eupolis, Timostratus, and Menander.

With the history of the Attic theatre is naturally connected the study of the choragic monuments, mementos of dramatic victories won at the festivals of Dionysus. It is well known that in Athens theatrical representations had the character of a national religious institution; the duty of forming and training the lyric choruses devolved upon the wealthiest citizens, who met the expenses of the *choragia*. The *choragus* had the duty of bringing together the various elements of the chorus, composed of men, boys, pyrrhic dancers, cyclic dancers, flute-players, and of furnishing the costumes, masks, and other accessories; he was selected according to tribe. If the chorus formed by him won the victory, the choragus received as prize a bronze

* Aristophanes, *Ran.* 297.

FIG. 27.—CHORAGIC MONUMENT OF LYSICRATES.

tripod, which he dedicated, in the name of his tribe, either alongside one of the streets of Athens termed Street of the Tripod, or in the temenos of Dionysus, or in the neighbourhood of the theatre. The tripod was placed on a small architectural monument, the character of which greatly varied. In a grotto, situated above the theatre, and transformed into a shrine by the victorious choragus Thrasyllus (320 B.C.), the tripods were placed on columns with triangular capitals. Other edifices assumed the form of a small pilastered temple, in the niche of which was placed the tripod received as a prize. The best known choragic monument is that of Lysicrates, in the Street of the Tripods. It is composed of a quadrangular base of white marble veined with blue, which supports a little edifice in the form of a rotunda. Six fluted columns with Corinthian capitals bear a frieze delicately ornamented with bas-reliefs, representing the defeat of the Tyrrhenian pirates by Dionysus, and their transformation into dolphins. The spaces between the columns are filled with marble slabs inserted with great care, and decorated in turn with tripods in relief. The edifice was covered with a marble roof in imitation of tiles, and was surmounted by a flower ornament, which was the direct support of the tripod. The date of this monument (335—334 B.C.), which offers one of the earliest examples of the Corinthian order as applied to architecture, is given in an inscription :—"Lysicrates, son of Lysitheides, of the deme Cicynna was choragus; the tribe Acamantis received the prize for the chorus of boys; Theon played the flute; Lysiades the Athenian trained the chorus; Euænetus was archon."

The name *odeon* was given to theatres especially designed for concerts; in these edifices poets and musicians exercised their talents before submitting their work to the judgment of the public. The odeons differed from the theatres chiefly in their roof and in several peculiarities of their internal arrangements; rows of columns supported the roof, and the seats, arranged in a semicircle, were very numerous. Athens possessed a celebrated odeon, erected by Pericles to take the place of the earlier building situated near the Enneacrunus. The odeon of Pericles was not far from the Acropolis; it adjoined the theatre of Dionysus on the east. The roof, of a pointed shape, was decorated with the masts and rigging of Persian ships, making the building resemble the tent of Xerxes. Plutarch speaks of the great number of seats and of columns which the building contained. Although it is difficult at the present time to form an exact idea of it, we may discover the general arrangements, common to this class of buildings, in the ruins of the odeon built by Herodes Atticus in memory of his wife Regilla. This was the largest of odeons known, and its ceiling of cedar was of rare magnificence. To-day, thanks to the excavations made in 1857 and 1858, we may clearly distinguish the tiers of seats with their corridors, and the walls, pierced with arcades, which supported the proscenium.*

* There is a restoration of the Odeon of Regilla, by M. Daumet, in the École des Beaux-Arts in Paris.

H

The *stadium*, where pugilists, runners, etc., con-
tended, was distinct from the *hippodromos* reserved for
chariot races. Their general arrangements differed but
slightly, because at the outset a very simple plan had
been adopted for each. They were large, level spaces,
enclosed within artificial embankments, or situated
between natural elevations, along which the spectators
assembled. The work of the architect and of the
engineer was limited entirely to the arrangement of
the barriers or starting posts (*carceres*). Those of the
hippodrome of Olympia, constructed by Cleœtas,
were famous ; they were, so to speak, boxes in which
the chariots were shut in ; when the gates, which
were placed along a curved line, were opened, it was
possible for the chariots to start at a signal without
giving any one of them an advantage over the others.
The arena was divided into two parts by an em-
bankment of earth, at the end of which was the limit,
around which it was necessary for the chariots to turn.

The stadium is a long and narrow arena, terminating at one of its extremities in a semicircle
(σφενδόνη), which resembles a theatre in its tiers of
seats and its rounded form. These tiers were extended along the sides of the arena, as may be seen in
the well-preserved stadium at Cibyra, in Asia Minor.
The essential difference between the stadium and the
hippodrome lies principally in the absence in the
former of the stalls and carceres intended to hold the
chariots ; in other respects we find the same arrangements, and analogous parts in each are designated by
the same names. With the progress of luxury, both

public and private, the stadia received greater elegance, as in tiers of marble seats, in colonnades
traces of which have been found in the ruins of the
stadium of Messene. The Panathenaic stadium in
Athens, used at the Panathenaic festivals, was magni-
ficently ornamented by Herodes Atticus, who faced the
seats with Pentelic marble. Pausanias and Philostratus speak with admiration of this expensive work,
executed in four years, which thoroughly transformed
the ancient stadium built under Lycurgus in 350 B.C.
It was here, many years later, that Hadrian caused
the Athenians to witness bloody gladiatorial contests,
which he introduced according to Roman usage.

Book Third.

SCULPTURE.

GENERAL WORKS.

K. O. MÜLLER, *Handbuch der Archäologie der Kunst*, 1830 (third edition by Welcker, 1845; English translation by Leitch, second edition, 1850).

BRUNN, *Geschichte der griechischen Künstler*, 1853—1859.

OVERBECK, *Geschichte der griechischen Plastik*; third edition, 1881—1882. *Die antiken Schriftquellen zur Geschichte der Bildenden Künste bei den Griechen* 1868.

L. M. MITCHELL, *A History of Ancient Sculpture*, 1883; *Selections from Ancient Sculpture* (phototypes), 1883.

A. S. MURRAY, *A History of Greek Sculpture*, 1880—1883.

COLLECTIONS OF ILLUSTRATIONS.

MÜLLER-WIESELER, *Denkmäler der alten Kunst*, 1854—1881.

CLARAC, *Musée de Sculpture*, 1826—1833.

WELCKER, *Alte Denkmäler*, 1849—1864.

O. RAYET, *Monuments de l'Art antique*, 1881—1884.

The great periodical publications, such as *Monumenti inediti dell' Instituto di Corrispondenza Archeologica di Roma*, 1829 *ff.*; *Archæologische Zeitung*, 1843 *ff.*; *Gazette archéologique*, 1875 *ff.*; *Journal of Hellenic Studies*, 1880 *ff.* etc.

Other important special works will be cited as occasion may demand.

CHAPTER I.

FIRST PERIOD.

LEGENDARY BEGINNINGS, DOWN TO THE CLOSE OF THE SEVENTH CENTURY B.C.

ALTHOUGH the industrial arts were early cultivated in Greece, the art of sculpture, properly so called—the art that attempts to reproduce the human form and to translate, as it were, religious or poetic conceptions into visible shapes—remained for a long time in a state of imperfect development. It was essentially religious at its beginning, and its first efforts were devoted to presenting the gods in tangible forms. This, however, could not be done until poetry, less trammelled, had given to the gods a fixed plastic character, and had thus furnished sculpture with the elements for its task. The development of sculpture was thus sub-ordinated to that of the poetic sense and religious sentiment.

The most ancient representations of Hellenic divinities were of the nature of symbols rather than of plastic works. The piety of primitive ages was content with a mere external sign which personified the god but was not an image of him. Thus Eros at Thespiæ, and the Charites at Orcho-menus, were symbolised by stones (ἀργοὶ λίθοι); Apollo Agyieus was represented by a column on the coins of Ambracia; the Hera of Argos was a simple column (κίων); the Dioscuri of Sparta had the form of two upright bars united by a cross beam. These rude symbols need not detain us; they belong rather to the field of illustrated mythology than to that of the history of Sculpture.

FIG. 28.
APOLLO AGYIEUS.
(On an Ambra-cian coin.)

The first statues in which the attempt was made to represent the gods in human form were of wood (ξόανα). These figures were clad in a garment as

tightly fitting as a sheath, "with shut eyes, arms
hanging down almost glued to the sides."* These
ancient wooden statues were without exception objects of worship. The Greek ascribed to them a
fabulous origin, and preserved them with religious
care. Pausanias, who lived in the time of Hadrian,
saw them in the temples, where they excited the
reverent emotion of devotees and the raillery of scep-
tics.† Kept carefully in repair, painted white or ver-
milion, they were clothed in rich materials, and owned
complete wardrobes like that of the Brauronian
Artemis, the contents of which are given us in an
inscription. Pictures of these ξόava are often found
upon painted vases ; but there are other materials for
study which give us more exact knowledge as to
their style. We have already mentioned the Artemis
of Delos as one of the most ancient monuments of
Greek statuary art. Without going back to the
remotest epoch, we are enabled from this marble from
Delos to understand something of the stiffness of
attitude of the ξόava. There is also in the British
Museum an imitation in stone of a ξόavov found near
Rhamnus in Attica. In these oldest wooden statues
the representation of the human face was most
infantile ; a simple line indicated the eyes, which
appeared closed. It is thus that the human figure is

---

* Diodorus Siculus, iv. 76.

† In 39 A.D. they figured in ceremonies of worship. Compare an
nscription from Cyzicus, mentioning the festivals celebrated in honour
of Antonia Tryphœna. Curtius, *Monatsberichte der Berlinere Akad. d.
Wissensch.*, January, 1874.

treated in some interesting marbles in the Louvre, one of which came from Paros. We here recognise without difficulty the vague and uncertain attempts of a rude and cautious art, which limits itself simply to reproducing the essential features of the face.

The Greeks ascribed their first progress in sculpture to Dædalus, a legendary personage said to have been born in Athens. " The statues he made were like living beings : they saw, they walked. It was he who first opened their eyes, unbound their legs and their arms."* By a tendency natural to their genius, the Greeks personified in him the achievements of a long period ; they never mention another artist with him. The name of Dædalus covers a whole school of obscure artists, who by a gradual progress had given some appearance of life and movement to the primitive images, and had emancipated themselves from a rigid tradition. Wooden statues ascribed to Dædalus were numerous, and Pausanias notes them carefully. They are distinguished from more ancient images by their open eyes, by their arms detached from the body, and by their attitude, which was that of a person walking.

This is not the place to dwell at length upon these modest beginnings of plastic art, nor is it necessary to dilate upon the fables in which the Greeks disguised these beginnings. In the view of the Greeks, all their arts were born on their own soil, and with each invention some marvellous or charming legend was connected.

---

* Diodorus Siculus, *ibid.*

The story of the potter Butades, who discovered the art of modelling, is well known. One evening his young daughter, the Corinthian Core, was struck with the idea of tracing upon the wall the silhouette of her lover, which had been cast there by the light of the lamp : her father filled out the profiles with clay, and the art of modelling was invented.* Separating the truth from the fiction in these fables, we find evidence to believe that the art of sculpture was first developed in Oriental Greece, as were all the arts of the Greeks. The technical processes were handed down in the families of artisans, and were gradually improved, until the day arrived when the modest blacksmith, who had forged armour, articles of furniture, and the like, became a cunning worker in metal, and applied to statuary art the technical knowledge and skill he had acquired. It is then that we find historical names, personages, and schools with distinct characteristics. Art emerges from its legendary stage, and by the aid of written records we are enabled to trace its subsequent history.

* Pliny, *N.H.*, XXXV., 151 ; Athenagoras, *Legat. ᴆro Christianis*, 14.

CHAPTER II.

SECOND PERIOD.

THE PRIMITIVE MASTERS, FROM THE END OF THE SEVENTH CENTURY B.C. TO THE SIXTIETH OLYMPIAD (540 B.C.).*

---

§ I. ORIENTAL GREECE.

SAMIAN masters excelled principally in metal working. They were toreuticians (workers in repoussé) and gold-smiths, and executed works then in demand among the princes of Asia. They did not limit themselves to specialities ; their activity showed itself in sculpture, and in all the other arts as well. Rhœcus, head of the Samian school, who lived about the fortieth Olympiad (620 B.C.), made a statue of Night, which was preserved in Ephesus. Theodorus, his son, pos-sessed wonderful skill, to which his statue of himself, holding a quadriga covered by the wings of a fly, bears witness. To the latter was also ascribed a statue of Pythian Apollo, executed for the Samians, in conjunction with his brother Telecles. These three artists represent the Samian school at about the fortieth Olympiad (620 B.C.) ; they were united not only in their labours, but likewise in tradition,

* *Cf.* Beulé, *Histoire de l'Art grec avant Périclès*, 1868 ; A. S. Murray, *History*, etc., Vol. I.

which ascribes to Rhœcus and Theodorus the inven
tion of casting in bronze, *i.e.*, the art of running the
molten bronze around a core of plaster.* Nothing
can give us a better conception of these early
schools of artists where the processes of the art were
transmitted from generation to generation.

The school of Chios also numbered within its list
toreuticians and sculptors in marble. Glaucus, without doubt a pupil of the Samians, was the maker of a
crater dedicated at Delphi by the Lydian king
Alyattes (about the fortieth Olympiad, 620 B.C.). He
seems to have kept up the tradition of working in
metals, and is credited with the invention of soldering
Towards the close of the seventh century B.C., however,
the Chian school of sculptors in marble had its
representative in Melas, who was followed by his son
Micciades, and by his grandson Archermus. M.
Homolle discovered at Delos a marble statue, which
gives us a clear idea of the style of these primitive
masters. It seems to belong to the first part of the
sixth century B.C. It is of a female figure, perhaps a
Victory, with raised wings, wearing a crown ; she is
represented in a very naïve manner, as running. An
inscription on the pedestal indicates it as the work
of Micciades and his son Archermus.† The works of
Bupalus and Athenis, sons of Archermus—about the
sixtieth Olympiad (540 B.C.), according to Pliny—were
of such artistic merit as to be transported to Rome

---

* Ancient bronze statues were made of separate pieces inlaid, as it
were, and riveted in place.

† *Bull. de Corr. hellén.*, 1879, plates VI., VII.

at the time of the conquest of Greece. At Chios was to be seen a statue by them of Artemis, which appeared sad to those entering the temple, but of a cheerful expression to those departing from it. At Naxos, at about the same time, Byzes, son of Euergus, a famous sculptor, executed statues for his country, and invented the marble tiling used in covering the joints in the roofing of temples.

These facts are enough to prove that the first development of sculpture was effected in Oriental Greece ; knowledge and technical skill carried thence into European Greece gave birth to those important schools which, between the fortieth and fiftieth Olympiads (620 B.C.—580 B.C.), conducted the art of sculpture to a comparatively advanced stage.

§ 2. EUROPEAN GREECE.

Theodorus the Samian had already been called to Sparta, there to erect the *Skias,* a metallic edifice in the form of a tent. The Æginetan Smilis had been trained in the school of Samos, and had made for the Samians a statue of Hera, which is figured on archaic coins of the island of Samos. But one fact above all others influenced the development of the art of sculpture in Peloponnesus ; this was the arrival of the Cretan masters, Dipœnus and Scyllis, at Sicyon, where they established themselves, and gave art a decided impulse (Olympiad L., 580 B.C.) ; their works were so far superior to those of native artists, that ancient writers ascribed to them the real beginnings of sculpture.

The Cretans wrought in bronze, in wood, and in Parian marble; their statues were in demand in the whole of Dorian Greece, in Ambracia, in Argos, in Cleonæ. At Sicyon were to be seen statues of the gods, works from their studios. Under the impulse given by them art spread throughout all Peloponnesus; many celebrated pupils are assigned to them, Hegylus and Theocles, the Laconian Dontas and Dorycleidas, Clearchus of Rhegium, Tectæus and Angelion, who wrought for the Delians, and together executed a statue of Apollo and the Three Graces or Charites, figured on an Attic coin.*

At about the same time art was developing at Sparta with great energy. By reason of the sojourn of Theodorus in this city, the art of working in bronze had made great progress there at the opening of the sixth century B.C., and the traditions of Samos had been appropriated by skilled masters like Syadras and Chartas. It was at Sparta that the disciples of Cretan masters were most numerous; but the Laconian school owed much to the Magnesian Bathycles, who established himself at Sparta after the year 544 B.C. (Ol. LIX. 1), bringing with him a group of Magnesian sculptors. Bathycles executed for the rude image of Apollo of Amyclæ a magnificent throne ornamented with gold and ivory, on which were portrayed in rich reliefs the principal Hellenic myths.

At the period of which we now write, Attic sculpture had hardly come into being. Lagging far behind

---

* Beulé, *Monnaies d'Athènes,* 1858, p. 364.

Peloponnesian art in point of time, it had not emerged
from the legendary stage until between the fortieth
and the fiftieth Olympiads (620 B.C.—580 B.C.). It
was soon after this time that Peisistratus beautified
Athens, erected the Olympieum, the older Parthenon,
and the earlier Propylæa. This artistic movement
to which Athens owed its earlier structures could not
have been made without plastic art playing some part
in it. It is at this date, as the farthest point, that
we must place the beginnings of the earlier Attic
school, the activities of which several years later were
to be felt in every direction. Before the sixtieth
Olympiad (540 B.C.) no artists are to be found in
Attica, except those of a legendary character, like
Simmias, who made a statue of Dionysus Morychus,
which at the time of the vintage was smeared with
lees.*

One fact of great importance contributed to hasten
the development of sculpture at about the sixtieth
Olympiad (540 B.C.) ; this was the frequency of portraits of athletes. Up to this time sculptors had
chiefly represented divinities, whose forms had been
fixed by a sort of hieratic conventionality. The
earlier statues of athletes, victorious in Greek religious
festivals, are marked by the traditional stiffness.
Pausanias describes the statue of the pancratiast
Arrachion, victor at Phigalia about the fifty-third
Olympiad (568 B.C.), in the following terms :—" The

---

* Endœus, of whom we shall speak later on, belongs likewise to the
last years of the first period, and not to the second period.

feet are barely separated, the hands fall at the sides and reach as far as the thighs."* The first victors at the Olympic games to receive statues at Olympia are Praxidamas of Ægina (fifty-ninth Olympiad, 544 B.C.), and Rhexibius the Opuntian (sixty-first Olympiad, 536 B.C.) ; these wooden statues cannot have been likenesses, but were symbolic images with no attempt at portraiture. Statues of gods and of athletes, treated in a conventional style, were for a long time the only attempts at representing the human figure. After the sixtieth Olympiad (540 B.C.) the figures of athletes were rapidly multiplied, and art thereby made a signal gain. Artists learned anatomy by copying the vigorous forms before their eyes ; they aimed to render accurately the structure of the body, the swell of the muscle, variety of attitude, and thus sought to realise the very essence of art, to achieve the perfect imitation of nature in life.

§ 3. MONUMENTS PRESERVED.

The monuments preserved from this period are few in number, and it is difficult to assign dates to them. They permit us none the less, however, to appreciate in them the progress that was being made, while their defects, by reason of the naïveté of unsuccessful effort, show us the conditions and the aims of the primitive masters.

The Artemis of Delos, which seems to be of the seventh century B.C., belongs also to the class of

---

* Pausanias, VIII. 40, 1.

ξόava.*   In the metopes of the temple of Selinus we
are in the last years of the seventh, if not at the
beginning of the sixth, century, B.C.†   These two bas-
reliefs of calcareous tufa were discovered in 1823.
One of them represents Heracles carrying off the
Cercopes ; the other Perseus, aided by Athena, killing
the Gorgon. They well characterise the style of
sculpture at the beginning of the period already discussed; the forms are heavy and massive, the muscles
are exaggerated, and the figures in general, the eyes
of which are large and obliquely set, betray the processes of primitive art.   The sculptor has emphasised
the ugliness of expression in the face of the Gorgon,
which is one of those horrible figures in favour with
the earlier artists.   Although of Greek workmanship, it recalls by its short and heavy proportions
Assyrian bas-reliefs.   These sculptures were painted
in accordance with an old usage, which seems to have
continued throughout the whole archaic period.

Some interesting bas-reliefs, now in the museum of
Sparta, which were found in the outskirts of the town,
are treated in an analogous way.   Two of these reliefs
fill up the principal faces of a rectangular monument ;
one of them represents Orestes meeting his sister
Electra, the other the murder of Clytemnestra by
her son.   The figures are short and heavily built ;
the movements indicated are awkward ; the attitude
of Orestes is almost identical in the two scenes.   We

* Homolle, *Bull. de Corr. hellén.*, 1879, plates II., III.
† Benndorf, *Die Metopen von Selinunt*, 1873. The date of these
sculptures may be set at about the fiftieth Olympiad (580 B.C.).

feel that the sculptor's chisel lacked the flexibility of

FIG. 29.—PERSEUS SLAYING THE GORGON.
(Metope from Selinus.)

action necessary to secure variety in the movements

I

of his figures. Several other bas-reliefs in the same museum, found at Chrysapha, represent two persons, a man and a woman, seated upon richly ornamented chairs : they are receiving the offerings of a cock and a flower which are brought to them by two very small figures. The subject represented has been a matter of controversy :* in the larger figures have been seen Asclepius and Hygieia, or, as is more probable, deified ancestors, conceived as divinities of the lower world, who are receiving offerings from their survivors. Whatever may be the subject, these old Laconian sculptures are of a very peculiar style ; the rigid outlines, which are stiff and accented, indicate a very primitive technique, which treated marble as if it were wood.

It would be interesting were we able to ascribe to fixed dates and definite schools some of these works. Perhaps we are to assign to the school of Dipœnus and Scyllis some fragments discovered at Actium, now in the Louvre. We should thus have evidence of the progress achieved under these masters. Though the foregoing is only a conjecture, these statues have their definite place in a series of interesting monuments that indicate the progress made in sculpture about the middle of the sixth century B.C. in the study of the human form, as found in the bodies of athletes. We refer to archaic statues representing Apollo. The most ancient of these appears to be the Apollo

* Von Sallet, *Zeitschrift für Numismatik*, 1882, p. 171 ; *Mittheil. des aeutsch. Inst. zu Athen*, 1877, p. 459, and 1882, p. 163. *Cf.* also Percy Gardner, *Fortnightly Review*, June, 1885.

of Thera, the aspect of which
accords in every particular
with the description given by
Pausanias of the statue of the
athlete Arrachion. Here is
quite the same attitude : the
arms fitting close to the body,
the legs barely separated, the
detail of the muscles but
feebly indicated, and on the
face a stupid smiling expression. It is safe to put the
date of this statue at about
560 B.C. The Apollo of Orchomenus differs but slightly
from that of Thera. The
attitude is the same, but we
here see the beginning of an
attempt to indicate more
energetically the muscular
development of the chest.
Next to this come the
statues in the Louvre, and
mark a new stage of pro-
gress.* About this time the
multiplication of portraits of

Statues of the same type have
lately been discovered in the excavations on the site of the temple of
Apollo Ptoüs at Akraiphia in Bœotia
by the French School of Athens.

I 2

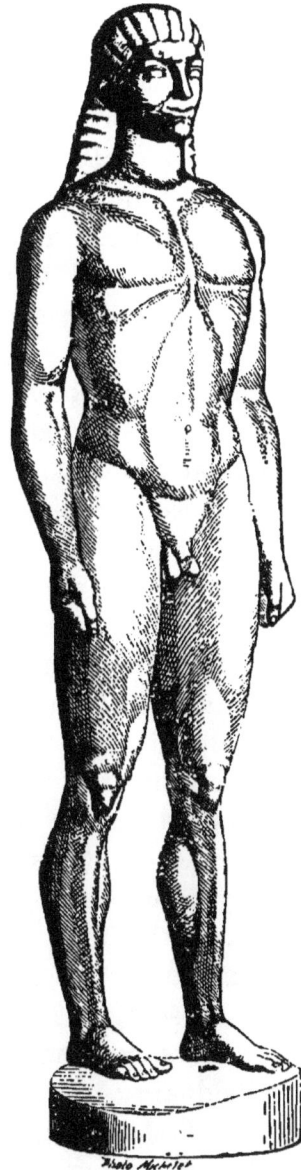

FIG. 30.
APOLLO OF TENEA.
(Munich.

athletes led Greek sculptors to study anatomy more
closely, and the results of these studies appear in the
statue of Apollo of Tenea, now in Munich. A greater
care in anatomical detail, more slender proportions,
a certain hollowing out between the body and the
arm, indicate a more advanced stage in art ; but the
attitude is still rigid, the face has that perpetually
simpering expression of archaic statues, with no individual character. Progress is shown principally in the
forms of the body ; the intelligent treatment of the
face required a mastery of art not attained by these
early artists.

These works belong to the Dorian schools. To
the Ionian, however, are to be assigned the ten
statues which decorated the avenue of Branchidæ at
Didymi, and are now, in part, in the Lycian room of
the British Museum. The inscriptions upon them
fix their date in the neighbourhood of the sixtieth
Olympiad (540 B.C.)—very valuable testimony when
we consider the uncertainty which too often prevails
in regard to the precise date of Greek monuments.
The statues, at first sight, suggest the Assyrian style.
They represent figures seated in chairs with high
backs ; the legs are close together, the hands rest
upon the knees. Such is the uniformity of treatment,
that these figures seem to have been executed after a
rule or *canon*, where no scope was left for the artist's
individuality.

At first glance, these statues from Branchidæ seem
to have nothing Greek about them. This is because
the Greeks dwelling upon the coast of Asia Minor were

in direct communication with the people of Phrygia and of Lydia, who served as intermediaries between Assyria and the Hellenes of Asia. All the monuments of art of this period, few though they are, found in Asiatic Greece, show a very pronounced Oriental influence. The museum of the Louvre possesses a

FIG. 31.—STATUES FROM THE SACRED WAY OF BRANCHIDÆ.
(British Museum.)

portion of a peculiar frieze from the old Doric temple at Assos, in the Troad. It exhibits a singular mixture of Greek subjects, with decorative motives borrowed from the East. Side by side with the contest between Heracles and the Triton we find zones of fighting animals, common in Assyrian art.* Again, to this period belongs the bas-relief in the Louvre, found

---

* Other fragments of the frieze were recovered in the excavations recently conducted by the Archæological Institute of America.

in the island of Samothrace, representing Agamemnon
seated upon a royal throne, accompanied by Epeius
and the herald Talthybius. While the figures are in
profile, the eyes are drawn as if facing the spectator,
and the hair is arranged in symmetrical curls.   The
figures are designated by inscriptions, the characters

FIG. 32.—BAS-RELIEF FROM SAMOTHRACE.

in which indicate a date near the sixtieth Olympiad
(540 B.C.).

This rapid survey of existing monuments reveals
anew the antagonism between the Dorian and Ionian
genius, which was to become yet more distinct. On
the one hand we see the Dorian genius, less directly
subject, it seems, to the influences of the Orient,

instinct with vigour and force; on the other, the Ionian genius, instructed in the school of Asia. The art of this period has, however, many features that are common to the two branches. It was, above all, religious, and its creations were such as were demanded by the requirements of worship and religious cultus; but even this condition was destined to be one cause of its gradual growth towards perfection. Religion was not a restraining influence in art; it did not impose upon it, as some have been tempted to believe, inflexible forms. On the contrary, the progress of religious conceptions cannot be separated from that of the conception of beauty; for the more beautiful a work of art became, the more welcome was it to the divinity. If we observe, in this primitive period, forms with slight variations often repeated, we must admit that this is the work of art alone, in the midst of the limitations under which it was developed. Followed principally in families of artists, in restricted schools, it comprised at first hardly more than a knowledge of technical processes: the master transmitted this knowledge to his pupil, who strove to extend it, until, finally, art grew bolder, and attempted the study of nature, which is the source of all real progress. After the period of the primitive artists, we enter upon that of archaic sculpture, which in turn touches upon the period of perfection. Greek art has found its path, and it strengthens itself by a vigorous and scientific study of the laws of progress in sculpture.

CHAPTER III.

THIRD PERIOD.

ARCHAIC SCULPTURE : FROM THE SIXTIETH TO THE EIGHTIETH OLYMPIAD (540 B.C.—460 B.C.).

---

§ I. THE DORIAN SCHOOLS IN PELOPONNESUS.

LESS than a century separates the period of primitive art from that in which sculpture attained its perfection. This is the time of those artists who, without being entirely released from the influence of archaic traditions, were preparing for the age of Pheidias and Polycleitus. The influence of Dipœnus and Scyllis caused the development in Peloponnesus of the various schools of Sparta, Sicyon, Argos, and Corinth, at the same time that the schools of Ægina and of Athens were affording evidence of energetic activity.

The most brilliant representative of the Spartan school, towards the close of the sixth century B.C., is Gitiades (Olympiad LXV.—LXXI., 520 B.C.—496 B.C.), who constructed for Athena a shrine adorned with reliefs in bronze. At the same time architect and sculptor, he regulated the use of the orders in the edifice, and planned the whole of the sculptural decoration of the temple, the subjects of which are drawn from Dorian myths. Elis, which was not far from Laconia, became the museum of all Greece, and was enriched by sculptures which accumulated in the *Altis*, the sacred grove of Olympia. Only one Elean artist, however, is known in the sixth century B.C. ; he was Callon, author of a group of thirty-seven figures in bronze, executed for the Messenians.

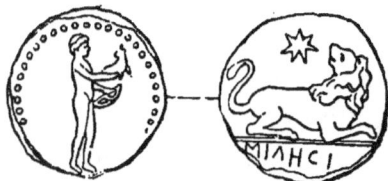

FIG. 33.—DIDYMÆAN APOLLO.
(On a coin of Miletus, when a free city.)

Greek writers do not mention the name of a single immediate pupil of Dipœnus and Scyllis in Sicyon, but we can easily see in this city the perpetuation of Cretan traditions ; Sicyon remained for a long time a veritable laboratory for the arts of metal-work-ing.* About the sixty-fourth Olympiad (524 B.C.) another Cretan, Aristocles of Cydonia, established him-self here, and his descend-ants worked at their art through the whole of the period now under discussion. Clecœtas, his son, invented for Olympia a system of barriers for the race-course, and executed for the Acropolis at Athens a statue of a warrior, coated

FIG. 34.
IMPERIAL COIN OF MILETUS.

* Pliny, *N.H.* xxxvi. 9.

with silver.   Aristocles, son of Cleœtas, formed a
school, and had many pupils, among whom were
Synnoön of Ægina and Pantias of Chios.

But the most famous of the masters in Sicyon was
Canachus, brother of the second Aristocles.   Among
his works are mentioned a seated Aphrodite, of gold
and ivory ; one of the Muses in a group made by him
in company with Ageladas and Aristocles ; above
all, two statues, one of Apollo of Didymi near
Miletus, the other of Ismenian Apollo, at Thebes,
resembling each other exactly, since these early
masters often repeated themselves.   The Apollo of
Didymi, executed between 494 B.C. and 479 B.C., was
a standing figure, holding a bow in the left hand and
a stag in the right.   The god is thus represented on
many coins of Miletus, where, in spite of dimensions
very much reduced, we may easily recognise an imitation of the Apollo of Canachus.   We find with
more certainty the principal features of the statue
preserved in bronze statues now in London and in
Paris ; the two most important are the Payne-Knight
bronze in the British Museum, and the Apollo
Piombino in the Museum of the Louvre.   The former
statue represents the god with the attributes which
Canachus had given him ; his hair is in curls across
his forehead, and long locks fall down his shoulders.
The bronze in the Louvre, however, can give us a
still better conception of this work of the Sicyonian
master.   It is a painstaking copy of the Didymæan
Apollo, executed in the last years of the sixth century
B.C.   The forms are vigorous and carefully studied ;

one of the legs, stepping slightly forward, gives the statue a firm equilibrium ; the hair is arranged in the archaic manner, and forms a heavy mass on the back of the neck ; the eyes were of silver, the lips and the nipples of the breast plated with red copper. The characteristics of this statue confirm the criticism of the ancients on the style of Canachus, which still retained much hardness and stiffness of treatment. By this statue we may judge of the state of plastic art at the close of the sixth and the beginning of the fifth century B.C., at an age when this Sicyonian artist brilliantly represented one of the great schools of sculpture in European Greece.

At Argos, which was near Sicyon, flourished a school which, about the seventieth Olympiad (500 B.C.), included several famous masters, among whom were Eutelidas and Chrysothemis, workers in metal; they had been preceded by a generation of sculptors, as is proved by the following inscription, cut on one of their works at Olympia :—" Eutelidas and Chrysothemis, Argives, made these statues : they received their art from their predecessors." But the great name of the Argive school is that of Ageladas, the period of whose activity extended from 515 B.C. to 455 B.C. His work is known only through the ancient writers, but this testimony shows that he successfully undertook most complicated and difficult subjects ; such was the statue of Cleosthenes of Epidamnus, whom he represented with his charioteer and horses. He also executed a group of captives and of horses, dedicated at Delphi by the people of Tarentum. Statues of

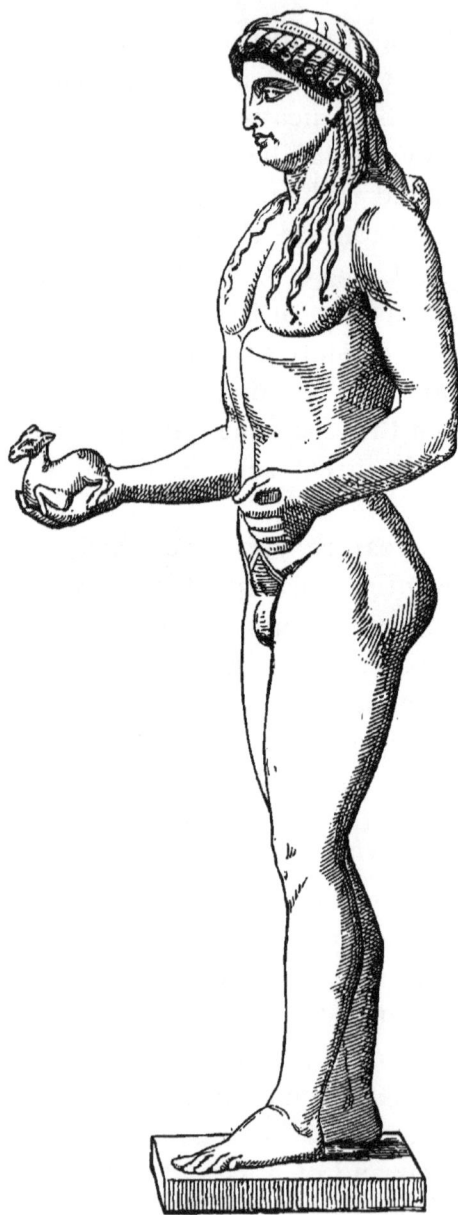

divinities from his hand are mentioned, such as that of Heracles Alexicacos ("averter of evil"), set up in the deme of Melite in Athens, and a statue of Zeus, for the Messenians. To Ageladas must therefore be assigned an important place in the history of Greek art, and to this distinction should be joined another not to be forgotten: this Argive sculptor was the teacher of Myron, of Polycleitus, and of Pheidias. By his instruction he prepared the way for the period of perfection, to which he nearly attained himself. Though no work of Ageladas has survived, we have reason to believe that his style greatly resembled that of the

FIG. 35.—PAYNE-KNIGHT BRONZE.   (British Museum.)

\FIG. 36.—HERACLES STRETCHING HIS BOW.
(Carapanos collection.)

Sicyonian masters.   Like them, he wrought in bronze, and, as is well known, through this material brought out precisely those qualities that are to be found in the archaic masters. A bronze statue must have perfect equilibrium, and other conditions of stability which demand most minute study.   It requires, further, great purity in its lines, minute and careful work in accessories, such as the hair, beard, etc.   It is probable that the masters of Argos and Sicyon had carried technical skill in modelling to its highest pitch, and that everything was now ready for the opening of a great age in the history of art.

But little is known of the Corinthian school of sculpture, to which belonged three artists, authors of a group dedicated at Delphi by the Phocians,* Diyllus, Amyclæus, and Chionis, who are known only from the mention made of them by Pausanias.   No marble has been found at Corinth that can give us certain knowledge of the methods and traditions of this school.†   We can hardly put confidence in vague indications pointing to Corinth as the source of a beautiful bas-relief belonging to the Carapanos collection.   A naked Heracles, his lion's skin thrown aside, is discharging an arrow ; all the muscles of his vigorous body are contracted by the effort, and his legs are stiffened by the movement which throws his body

---

* Pausanias, x. 13, 7.

† A round altar found at Corinth, now in England, representing the reconciliation of Heracles with the divinities of Delphi, is nothing more than a late imitation of the archaic style. On these imitations, which are archaistic rather than genuinely archaic, see below, p. 144.

forward. No name is affixed to this relief; but the general features of its style place it at the period at which sculptors copied nature, without being able to disengage themselves from servile imitation. (Fig. 36.)

We will not pass to the next school without specifying an interesting monument of Bœotian sculpture. This is a group in porous stone found at Tanagra. The monument was erected upon the tomb of two friends, Dermys and Citylus, by a person named Amphalces; the two friends are represented as embracing, each with his arm about the neck of his companion.* The proportions of the figures are slender; an effect of very marked elegance is produced, together with that of striking inexperience. The Bœotian artist was less advanced in the science of anatomy than were his brothers of Peloponnesus.

§ 2. THE ÆGINETAN SCHOOL.

The school of Ægina was connected with the Dorian schools; this is apparent from the marbles preserved, and the history of the Æginetan school clearly shows its Dorian origin. As early as between 550 B.C. and 536 B.C., Smilis, the founder of the Æginetan school, visited Elis and executed works for Olympia. About the seventieth Olympiad (500 B.C.), Callon of Ægina, a pupil of Dorian artists, exercised his art in Sparta and in Corinth, and his statues of gods were marked by the harshness characteristic of

* A. Dumont, *Gazette archéologique*, 1878.

Dorian works of art. Still other names are mentioned
by ancient writers, which bear witness to the activity
displayed in the studios of Ægina :—Glaucias,
Anaxagoras, Calliteles, Simon, Synnoön and
Ptolichus, Serambus, Theopropus, all of them skilful in
modelling the figures of athletes and of horses at the
order of Olympian victors, such as Gelon of Syracuse,
who desired to be represented with their quadrigæ.
The last member of the Æginetan school, which
did not survive the conquest of the island by the
Athenians, seems to have been Onatas, who flourished
after the Persian wars. Among the numerous statues
of divinities ascribed to this sculptor, mention is made
of an Apollo of Pergamum figured on copper coins of
the time of Marcus Aurelius.

It is Ægina herself that furnishes the most important groups of monuments for this whole period.
The sculptures which adorned the pediments of the
temple of Athena in Ægina, discovered in 1811 by
Danish, German, and English travellers, are the most
beautiful specimens of Greek art in the archaic age
now in existence.* The subjects are taken from the

---

* These marbles, obtained in 1812 by Prince Louis of Bavaria and
restored by Thorwaldsen, are in the Glyptothek of Munich. They have
often been published, notably by the French in *L'Expedition scientifique
Morée*, 1831—1838. For bibliography consult Brunn, *Beschreibung
der Glyptothek*, Third Edition, 1874. K. Lange has lately written
a work on the important question as to the arrangement of the
figures or the composition of the two pediments (K. Lange, *Die
Composition der Ægineten*, 1878). On studying the monuments with
extreme care, as had already been done by H. Prachov (*Annali
dell' Inst*, 1873), Lange shows that the composition of the two

Homeric poems, and are connected with the history of the heroes of Ægina. On the eastern pediment,

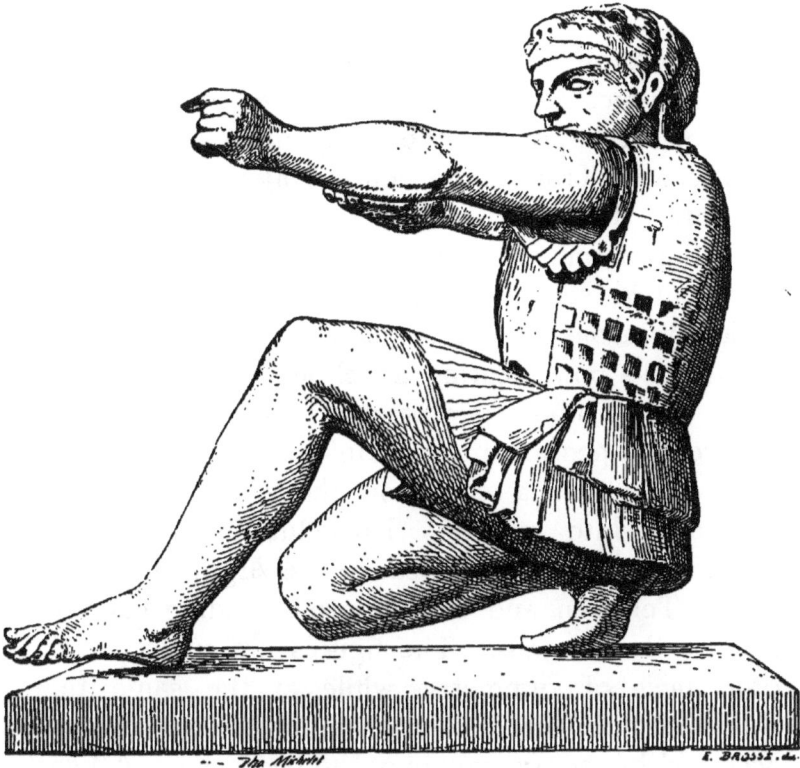

\Fig. 37.—HERACLES.
(From the Eastern Ægina pediment.)

pediments is rigorously symmetrical. Each scene had fourteen figures instead of eleven, as had been previously believed. In the centre, Athena, with the fallen warrior. On either side—(1) a figure bending towards the prostrate body; (2) two warriors, armed with spears, standing erect; (3) two warriors kneeling and fighting, one with a spear, the other with a bow; (4) finally, in the corner, a wounded man attempting to draw an arrow from his pierced breast.

J

of which five figures and some fragments are preserved, the scene is that of the episode of the conflict between Heracles and Telamon, son of Æacus, king of the island, on the one hand, and Laomedon on the other. At the feet of Athena, who occupies the centre of the pediment, lies the body of Oëcles, one of the companions of Heracles; Greeks and Trojans are contending over the fallen warrior. The small number of figures preserved does not permit the restoration of this first scene with entire certainty; but we recognise without difficulty in one of the forms Heracles kneeling and drawing his bow, wearing a helmet made like a lion's head, while his body is protected by a leathern cuirass. The second pediment, which in great measure can be restored, presents a scene from the *Iliad* (XVII. 715 *ff.*), translated into marble. Patroclus has just been vanquished; two heroes dear to the Æginetans, Ajax, son of Telamon, and Teucer, are defending his body against the Trojans. The figures are subject to the rigid laws of symmetry, while at the same time skilful composition has utilised the triangular space occupied by the tympanum of the pediment. Athena, standing erect and with a spear, presides over the combat; on the one side are Trojans, on the other Greeks, while in the front Ajax is defending his comrade, who has fallen at the feet of the goddess. Midway towards the corners, under the projecting coping of the pediment, the figures are kneeling; at the corners the wounded, in a lying posture, complete the picture, the lines of which conform to an absolute parallelism. Certain parts of the statues were coloured, and some of the accessories were executed in bronze. The art of these statues, though so skilful and so accurate from an anatomical point of view, is still under archaic influences; these appear principally in the modelling of the heads and of the hands and feet, and in the uniformity of types: the faces are expressionless, and are encircled by carefully curled hair; the lips are narrow and pursed up, and have the stupid simper of figures of the ancient style, while the hands and feet are treated awkwardly. But the attitudes and many minor details indicate a scrupulous observation of nature; in the vigorous forms we discover a treatment of muscles that is restrained, but at the same time of almost geometrical precision. The Æginetan artists possessed two essential qualifications—knowledge of the laws of both statics and anatomy. The ancients recognised their style by certain delicate characteristic features; we must, however, associate the school closely with the Dorian tradition. A genuine Dorian instinct reveals itself in the taste for the exact representation of nature, and in the vigorous execution which gives us firmly knit forms and bestows upon the Æginetan marbles an austere beauty.

§ 3. THE ATTIC SCHOOL.

At Athens the period of the rule of Peisistratus coincides with that of an early development in art that seems to have been fruitful. The great works of art executed at the order of Peisistratus contributed

J 2

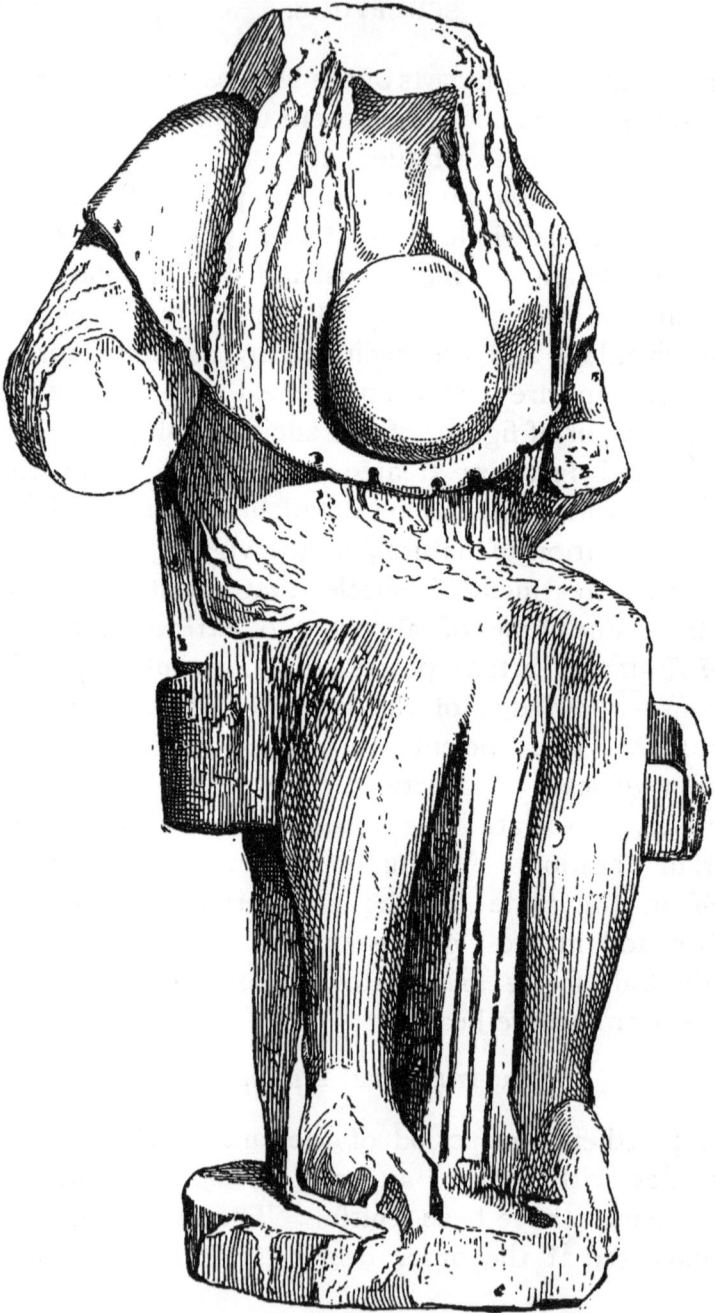

FIG. 38.—SEATED FIGURE: ATHENA. (Athens.)

to the rapid progress of sculpture.   At the beginning of this period—about the fifty-seventh Olympiad

FIG. 39.—HEAD OF AN ATHLETE.
(Rampin Collection.)

(552 B.C.)—we find the name of Endœus, said to have been a pupil of the legendary Dædalus.*   Endœus

* An inscription bearing the name of Endœus seems to date as far back as the sixtieth Olympiad (540 B.C.).—*Corpus Insc. Attic.*, I. 477.

was the author of a seated figure of Athena, dedicated by Callias in the Acropolis of Athens.   He
made a copy of this statue for the city of Erythræ.
We may recognise, if not the original, at least an imitation of this Athena in an Athenian marble now preserved on the Acropolis.* (Fig. 38.) As in the statues
at Branchidæ, the goddess is here seated in an attitude
at once rigid and hieratic.   Upon the ægis that covers
her breast fall the locks of her long hair symmetrically
parted.   The artist endeavoured, above all, carefully
to reproduce with great minuteness the fine folds of
the tunic, undulating along the full length of her form.
There is here an evident imitation of ancient wooden
statues, and of the folds of drapery that covered them.
With Gorgias, Aristion, Callonides, and Epistemon,
Endœus represents the first Attic school, contemporary with Peisistratus, which we are better able to
study by reason of the monuments preserved.   Of
these we will mention the more important.

First among these is the head of an athlete,†
which shows the progress made from the Apollo of
Orchomenus and from that of Tenea. (Fig. 39.) The
workmanship is very fine.   The athlete, with hair and
beard tightly curled, is crowned with an oak chaplet,
and, in spite of awkward execution, a successful attempt
at elegance is manifest.   Individuality, still only feebly
indicated in this work, is more clearly marked in two

* We know of a repetition of this statue of a date undoubtedly
more recent.

† A. Dumont, *Monuments grecs de l'Associat. des Études grecques*,
1878. (Collection Rampin.)

other heads, also from Athens. The former is that of
an athlete with an expression full of energy, whose
ears are misshapen from blows received in boxing.*
The artist has plainly tried to render the personality
of his model, and has succeeded in giving his face the

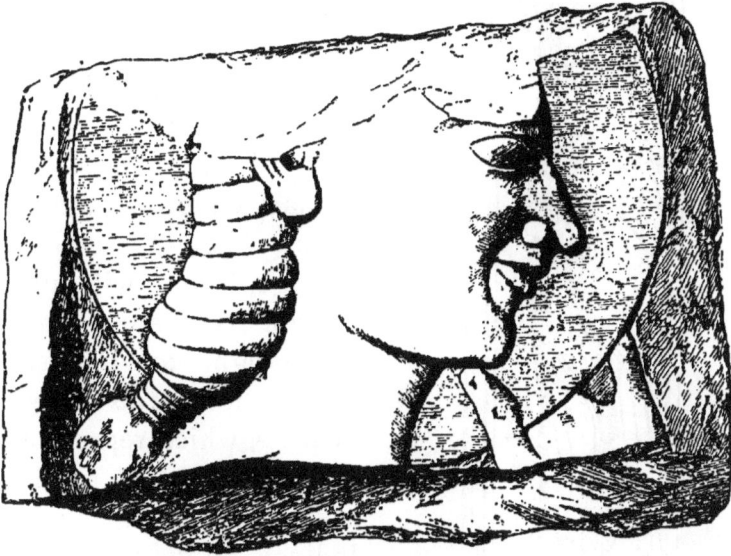

FIG. 40.—DISCOBOLUS.
(Museum of Athens.)

character of a portrait. Elegance is more evident in
this work than in another head in bas-relief representing a *discobolus*, which seems to have been part of
a *stele*, bearing the name of Xenophantus. (Fig. 40.)†
The features are fine, somewhat elongated, and were

* O. Rayet, *Monuments grecs*, etc., 1877.
† O. Rayet, *ibid.*

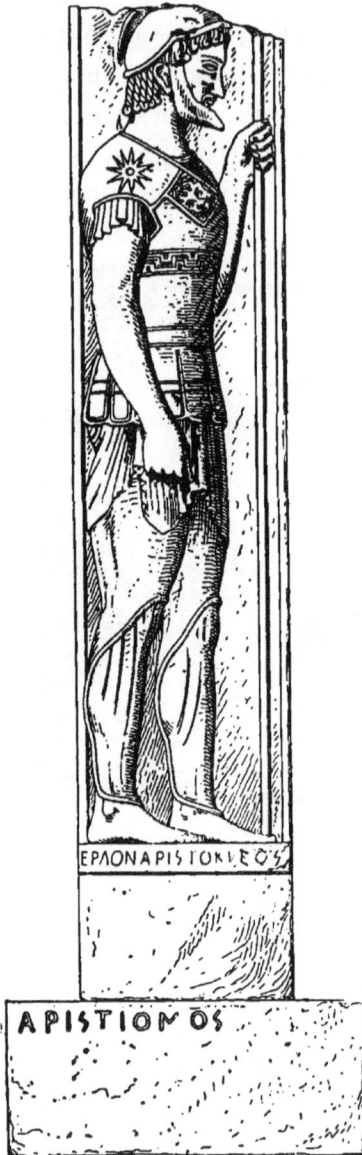

FIG. 41.—STELE OF ARISTION.
(Found at Velanideza.　Museum of
Athens.)

copied from life.　It is the true Athenian type, with all its characteristics.　The hair is twisted and tied together into a mass by a band: a kind of semi-Oriental coiffure which the Greeks abandoned after the Persian wars.　The eyes still retain that obliqueness given them by the archaic sculptors ; the total effect, though somewhat strange, is not without its charm.　We are probably to assign to the same period a statue now in Athens, which represents a bearded figure (perhaps Hermes) bearing a young bullock on his shoulders.

Among the first works of the Attic school are to be placed many *stelæ*, upon which, according to an ancient usage, were portrayed likenesses of the dead.　Such is the stele found at Velanideza in Attica, known by the incorrect name of the "Soldier of Marathon."

The inscription cut below the relief indicates that the monument is that of a person named Aristion ; * the author is the sculptor Aristocles, perhaps son of Aristocles the Cydonian, who figures among the masters of the school of Sicyon. The bas-relief,

A.   B.
FIG. 42.—A. THE TYRANNICIDES ON AN ATHENIAN COIN.
B. THE SAME, ENLARGED.

upon which traces of colouring are still visible, represents a Greek warrior in full armour ; his hair and beard, made into tight curls, were brownish ; his cuirass was dark blue, and the ground of the stele red. By means of the inscription, the letters of which seem to

* Ἔργον Ἀριστόκλεος ‖ Ἀριστίονος.

indicate the seventieth Olympiad (500 B.C.), we are able to group this monument with others which show us the first attempt of Attic sculpture.

After the banishment of the Peisistratidæ, the movement in art was not arrested. Names of well-known artists testify to this. Antenor was charged by order of the people to execute the statues of Harmodius and Aristogeiton, the slayers of Hipparchus: Amphicrates sculptured a lioness in memory of the courtesan Leæna, the brave accomplice of the two tyrannicides. Carried away by Xerxes at the time of the Persian wars, the group of the Tyrannicides was replaced by another in bronze, the work of the sculptors Critios and Nesiotes; Alexander, many years later, brought back from Asia the work of Antenor, and restored it to the Athenians. The group of Critios and Nesiotes is reproduced on an Athenian coin. Small as are the dimensions of this representation, if we regard it in conjunction with an analogous bas-relief upon the sides of a marble chair in Athens, we obtain a sufficiently clear idea of the original group of Critios and Nesiotes to enable us to see a copy of it in the group of the Tyrannicides in the museum of Naples. The two conspirators are represented as brandishing their daggers and moving forward at a rapid pace. The modern restorations removed, we see in this group traces of an already advanced style. In these two sculptors, with their contemporary Hegias, we touch closely upon the borders of the period of perfection. Hegias, of whose work we know only through the statements of ancient writers, was the first master of

FIG. 43.—WOMAN ENTERING A CHARIOT.
(Bas-relief on the Acropolis of Athens.)

Pheidias, before the great sculptor went to Argos to put himself under the instruction of Ageladas. Lucian compares the works of these masters with those of the earlier Attic orators ; he found them sinewy, somewhat harsh, of careful design, indicating labour and pains.*

This judgment of the rhetorician of Samosata may be applied to the sculptures preserved, contemporary with these last of the archaic masters. Such is a bas-relief discovered on the Acropolis representing a bearded person (Hermes or Theseus), wearing a petasus upon his head and clad in a tunic with small folds. Besides this there is the interesting bas-relief of the Acropolis, in which a woman is preparing to enter a chariot. (Fig. 43.) Is this Athena, or is it a Wingless Victory? No distinctive divine attribute marks this figure, which is draped in a long tunic and a robe, the folds of which are treated with the most delicate art. The proportions of the bas-relief are slender and fine ; while in its style we see at their dawning the special qualities most characteristic of Attic genius. †

Atticism is, in fact, one of the most persistent and permanent of the manifestations of Greek genius. Towards the close of the sixth century B.C., Athenian artists showed the same tendencies that we find two centuries later in Praxiteles : thorough know-

---

* Lucian, *Rhet. præcept.*, 9.

† All the monuments of the first Attic School cannot be mentioned here. *Cf.* Schöne, *Griechische Reliefs*, 1872, the plates of which represent some of these works.

ledge, taste for the exquisite, and a keen feeling for grace in art. Long, slender, elegant figures, at times too slender, that remind one of Florentine sculptures; great dexterity in the treatment of light drapery, which seems almost translucent; low relief; patient and minute care in workmanship—such are the leading characteristics of Attic sculpture at the beginning of the fifth century B.C.

§ 4. ASIATIC GREECE AND THE ISLANDS.

If we bear in mind that Attic genius was one of the varieties of Ionic genius, we shall not be surprised at finding, in part at least, its characteristics in the sculpture of the Ionians. The British Museum possesses some bas-reliefs found by Mr. Fellows (1838) at Xanthus, in Lycia; they came from an edifice known as the "Harpy Monument." The subjects depicted are symbolic scenes, in which are portrayed winged harpies, with human heads, carrying off in their talons small figures, the personifications of souls. At the first glance we are struck by the analogy subsisting between these marbles from Xanthus and Attic sculptures. Undoubtedly in the line of the same tradition may be placed a statue of Hera (now in the Louvre), found at Samos, belonging to the middle of the sixth century B.C.* (Fig. 44.) In spite of the stiffness of the main lines, the execution betrays a hand already sure. The artist has endeavoured to render the

* P. Girard, *Bulletin de Correspond. hellén.*, 1880, plates XIII., XIV.

lightness of tissue in the drapery. It is easy to detect in these works—all of which have a family resemblance—at least the germs of all those qualities that are peculiar to the Ionian genius.

The artistic movement, of which we have sketched the history, extended to all parts of the Greek world. This is proved by sculptures found outside of the countries already mentioned. The island of Thasos produced the bas-reliefs that were brought away by M. Miller, and are now in the Louvre. They were carved upon slabs of marble that may have formed the periphery of an altar, and represented Apollo leading the chorus of nymphs, and Hermes with one of the Graces (or Charites). An inscription leaves no mistake as to the meaning :—"To the Nymphs, and to Apollo Nymphegetes, sacrifice what victims you choose, male or female : sheep and hogs are forbidden : no pæan is sung."

The progress of art was felt even as far as Macedon. At Pharsalia M. Heuzey discovered the charming bas-relief styled by him the " Exaltation of the Flower." * Two young girls seem to be in conversation ; one of them presents a flower and a fruit to her companion, while the other, holding up in her right hand a full-blown flower, seems to exalt and worship it. With no wish to define too narrowly the meaning of this charming dialogue, Heuzey believes that this monument refers to the cultus of Core, daughter of Demeter, a divinity suggesting in Greek

---

* It is without doubt a fragment of a mortuary stele.

legend the ephemeral but incessantly recurring bloom of nature. While we may not be able to infer the existence of a distinct Pharsalian school of sculpture, it must be recognised that Greek art at this epoch had a great power of expansion. This bas - relief, shows that artistic activity had penetrated into the northern parts of Greece. It may, however, have been produced in the Ionian schools of Asia Minor ; for we cannot maintain with certainty that at this time Greece had any artistic schools that were distinct from those of the Ionians.*

The museums of Europe possess still other marbles that one is tempted to ascribe to the period which we are studying. But we must

* Cf. the theory of Brunn, *Paionios und die nordgriechische Kunst* (*Sitzungsbericht der bayer. Akad.*), 1876.

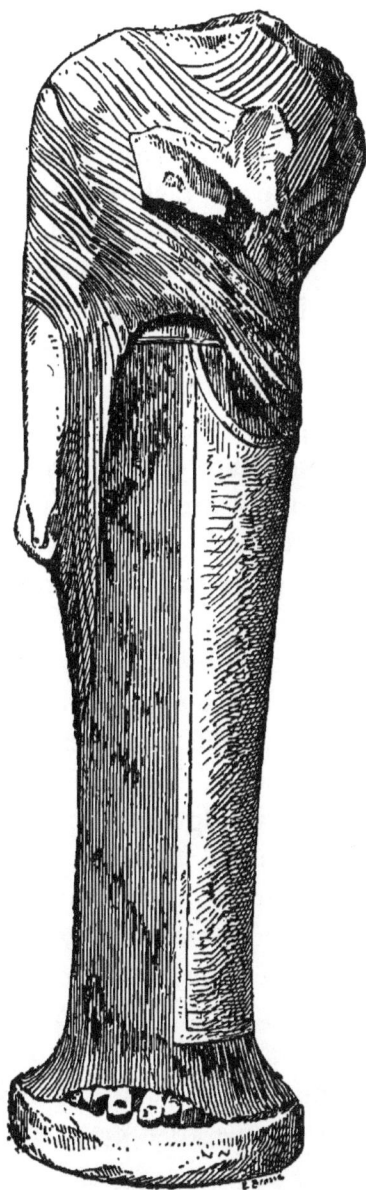

FIG. 44.—HERA.
(Statue found at Samos.)

distinguish with great care between really archaic and archaistic art, the latter being mere imitation. At certain epochs when the creative energy in art has become exhausted, men return, under the influence of fashion, to a pseudo-archaism, a sort of archæological reproduction of an ancient style.* Thus the altar of the twelve gods (Louvre) seems to be a copy of that raised by Peisistratus, and is not earlier than Hadrian's time. Likewise the archaistic Athena of Dresden, and the Artemis of Naples—the base adorned with reliefs representing the dispute of the Tripod, which is also to be seen in Naples—are works of an assumed archaism, and entirely lack that naïveté which is characteristic of original works. Imitations of this class, however, are of great interest ; they emphasise the essential notes of the style which they aim to reproduce.

We see by the monuments of this period that the archaic masters were possessed of great knowledge and skill ; the Dorians had a sternness of design and a knowledge of the human body that were not to be surpassed. They lacked, however, the power to rise superior to their knowledge; they needed to be emancipated from slavery to the mere study of the human model ; they had not yet that flexibility and liberty which constitute genius. Thus, in spite of their excellent qualities, their works are stiff and harsh ; energy and precision are carried to excess. Were an artist of genius to appear, he would put into practice this

* Cf. in Le Bas-Foucart, Inscr. du Péloponèse, p. 53, an archaistic statue executed 113 B.C.

great knowledge, he would animate it with his own personality; he would bring it into the service of larger conceptions, and art would thus attain perfection. It is one of the general laws of art, that great epochs are prepared for by the patient labour of earlier masters; there is no sudden bursting forth of genius.

These archaic masters are what in our day would be called realists. It was through lack of accurate knowledge of them that Winckelmann wrote:—"Given up to a blind study of the ideal, they wrought in accordance with a general system adopted by them, rather than after nature." On the contrary there is no nearer approach to nature in art than in the marbles of this period. The archaic masters copied from life; if we are to seek anywhere for the true Greek type, we can beyond doubt find it in the marbles of the earlier style, the very exaggerations of which are a pledge of truth. It is for these reasons that archaic Greek sculpture deserves a profound study. There is no keener pleasure than that of analysing the characteristics of genius at its struggling infancy, and of tracing these characteristics in the earnest and unconcealed efforts that lend to its creations the paramount charm of sincerity.

K

CHAPTER IV.

FOURTH PERIOD.

SCULPTURE FROM THE EIGHTIETH TO THE NINETY-SIXTH OLYMPIAD (460 B.C.—396 B.C.).

---

§ I. CALAMIS AND MYRON.

IT was in Attica and in Peloponnesus that sculpture attained its highest perfection during the brief but brilliant period between the Persian wars and the first years of the fourth century B.C. In Athens the name of Pheidias towered above all other names, and it has been customary to regard him as the uncontested head of the Attic school. At the same time we should not forget that at his side lived artists who still retained their own individuality and native bent. Greek art of the best period remained free and varied; it did not warp itself to suit the formulas of schools. The more we know of this wonderful fifth century B.C., which produced the greatest works of sculpture, the more are we convinced that perfect liberty characterised the activity of Greek art at its best.

In order to obtain a just idea of the Attic school in the second half of the fifth century B.C., we must first consider those masters who were still under the spell of ancient traditions, the most important of whom are

Calamis and Myron. The period of Calamis' greatest activity extends from the seventy-fifth to the eighty-seventh Olympiad (480 B.C.—432 B.C.) ; he was, then, a contemporary of Pheidias. Like the old masters that preceded him, he attempted every department of plastic art : groups of bronze, colossal statues, figures of gods, of men, of animals ; his work comprehended every variety. He excelled especially in animal figures, as testify his two racing horses (κέλητες ἵπποι), ridden by boys, which were greatly admired at Olympia. Pliny speaks of his colossal statue of Apollo at Apollonia, in Pontus, as a masterpiece of daring art. The ancients greatly praised a statue known as Sosandra, either an Aphrodite Pandemus or an Aphrodite dedicated by an Athenian near the entrance to the Acropolis. Lucian mentions its " grave and lurking smile,"* and the well-ordered and becoming arrangement of its drapery. There was but little harshness in the style of Calamis, which was compared in antiquity with that of the orator Lysias for its " delicacy and grace."† These are the excellencies and the defects of the earlier Attic school. Calamis executed for the people of Tanagra a Hermes Criophorus (carrying a ram upon his shoulders), which is reproduced on a bronze coin of the city. Guided by this reproduction, we recognise copies of the Hermes of Calamis in a terra-cotta statuette from Tanagra, and in a marble in the Pembroke collection.

The date at which Myron is to be placed is a

---

* Lucian, *Imagines*, 6. (Τὸ μειδίαμα σεμνὸν καὶ λεληθός.)
† Dionysius Halic., *De Isocr.* 3, p. 542.

K 2

matter of great controversy. Brunn * does not accept
the date given by Pliny (ninetieth Olympiad, 420 B.C.).
It is probable that Myron, who was born at Eleutheræ,
lived about the eightieth Olympiad (460 B.C.), and was
a contemporary of Calamis. Like Pheidias, he was a
pupil of the Argive Ageladas, and remained faithful
to old traditions by working almost entirely in
bronze. He made numerous statues of gods and of
athletes, many of which were carried off by the
Romans. His group of Marsyas and Athena,
where the satyr starts back in a surprised attitude,
was the inspiration of several similar monuments
preserved in our museums, among which are the
satyr of the museum of the Lateran, in Rome,
and a bronze in the British Museum.† A copy of
Myron's Discobolus is certainly recognisable in a
famous statue in Rome (at the palace *Massimi alle
Colonne*), which represents a player about to hurl
the discus. The ancients considered the Discobolus
the work of a daring innovator ; in reality the life-
like attitude of the player testifies that the artist had
entered upon a new method marked by freedom and
boldness of style. Myron excelled in his renderings
of material life ; his animal figures bore the stamp of
reality, and the poets of the Anthology praise to
excess a bronze cow executed by him. The epigram
ascribed to Anacreon is well known :—"Shepherd,
pasture thy flock further on, lest thinking thou seest

* Brunn, *Geschichte der griech. Künstler*, Vol. I., p. 242.

† *Cf.*, on the Marsyas of Myron, M. Collignon's notice in Rayet,
Monum. de l'Art antique.

FIG. 45.—COPY OF THE DISCOBOLUS.
(Rome.)

the cow of Myron breathing, thou mayest wish to
carry her away with thy oxen." *

With Calamis and Myron, another artist who did
not belong to the Attic school, Pythagoras of

FIG. 46.—METOPE OF THE TEMPLE OF THESEUS.
(Athens.)

Rhegium,† well represents that generation of sculptors
who may be termed the last precursors of the age of
perfected art.   In point of time, they already belong
to the period of perfection ; but their style still

* *Anthol. Palat.*, 715.
† *Cf.* Waldstein, *Pythagoras of Rhegium and the Early Athletic Statues,*
in *Journal of Hellenic Studies,* I. pp. 168—201, and II. pp. 332—351.

FIG. 47.—DEMETER, TRIPTOLEMUS, AND CORE.
(Bas-relief from Eleusis. Museum of Athens.)

exhibits the archaic character, which is now soon to be left behind. The period corresponds politically with the close of the influence of Cimon, when Athens was making good the devastation caused by the Persian wars, and was about to enter upon the most glorious epoch in her history.

The sculptures of the temple known as the Theseum, in Athens, show us the transition from the archaic style to-that of this epoch. Begun in the fourth year of the seventy-seventh Olympiad (469 B.C.), and finished after the eightieth Olympiad (460 B.C.), this temple, which is known as the Theseum, is an excellent specimen of the Doric order. Though nothing has been left of the pediments, eighteen metopes have been preserved. They portray a series of the labours of Heracles and of the exploits of Theseus. The frieze that runs around the wall of the cella represents the conflict at the marriage of Peirithoüs, between the Lapithæ and Athenians on the one hand, and the Centaurs on the other. The style is one of great energy, the attitudes are life-like; it is the composition alone, still cold and somewhat harsh, that indicates that these marbles are earlier than those of the Parthenon.

The same remarks apply to the beautiful bas-relief found at Eleusis, where Demeter, with her daughter, is giving to the youthful Triptolemus the grain of wheat which is to fructify the Rharian plains. This work is one of great beauty, and the delicacy of the modelling, the suppleness and grace of movement, are those of the purest style; but in

certain details are to be recognised the influences of
archaic art, still very persistent, from which Attic
sculpture did not emancipate itself until under the
spell of the genius of Pheidias.

§ 2. PHEIDIAS AND HIS SCHOOL.

Under the government of Pericles everything contributed to give a wonderful impulse to the arts :
national pride, exalted by the victories of the Persian
wars, the necessity for rebuilding edifices laid waste
by the enemy, together with the progress of an
enlightened democracy, gave to the public spirit
of Athens a remarkable freedom. " Athens,"
said Pericles, " abundantly provided with all the
means of defence required in war, ought to use
her wealth in works which, once finished, should
assure her an undying glory."* These words sum up
Pericles' system of administration. Thus, under the
government of this great statesman and orator, Athens
was adorned with edifices with a rapidity that long
amazed the Greeks. As late as the time of Plutarch
these monuments possessed all the " freshness and
brilliancy of youth, glowing with that charm of
novelty which preserved them from the touch of
time."† After the Theseum and the Temple of
Wingless Victory, were erected the Parthenon, finished
in the third year of the eighty-fifth Olympiad (438
B.C.) ; the Propylæa (between the eighty-fifth and

---

* Plutarch, *Pericles*, 12.   † Plutarch, *ibid.*, 13.

eighty-sixth Olympiads, 440 B.C.—436 B.C.), and the
Erechtheum, begun in the year before the opening of
the Peloponnesian war, and finished about 393 B.C.
These structures occasioned activity in all the arts,
and especially in sculpture. Sculpture, in fact, was not
separable from architecture : it sought no isolated
path. At that epoch of faith and belief its supreme
function was to aid in adorning the temples of the
gods; and since religion is, in reality, one of the
forms of public life, art in the time of Pericles received from the religious and from the national sentiment a dignity and a purity of inspiration that can
never be surpassed.

It is well known that Pericles entrusted Pheidias
with the direction of the artistic works. Pheidias was
born probably about the seventieth Olympiad (500
B.C.), and first went through a period of study under
Ageladas. The second period of his life was passed
under the administration of Cimon. He then executed
statues, the subjects of which related to the Persian
wars ; among others, a bronze group consecrated at
Delphi, made from a tithe of the booty captured at
Marathon. From this period also dates the colossal
bronze statue of Athena placed upon the Acropolis,
and known by the name of Athena Promachus or
Πυλαιμάχος. She stood with one hand holding her
lance, while the other held her shield, which, much
later, was decorated with bas-reliefs by Mys. From
afar could be seen the point of the spear and the crest
of her helmet towering high above the Acropolis.
Athenian coins, which show the Acropolis in profile,

give us some idea of the enormous dimensions of the statue. Here might be mentioned other works by Pheidias, the precise dates of which are uncertain: statues of gold and ivory for the most part, such as the Athena Lemnia of the Acropolis, an Amazon, and a bronze Athena later transported to Rome.

But the period of Pheidias's greatest works is that which co-incides with the administration of Pericles, when the great sculptor was surrounded by a group of artists working under his directions

FIG. 48.
THE ACROPOLIS.
(On an Athenian coin.

—Alcamenes, Agoracritus, Cresilas, Colotes, Pæonius, and his own brother, the painter Panænus. It was at this time that he executed the Athena Parthenus and the Zeus of Olympia.

Concerning the close of his life there is but little certainty, on account of the contradictory stories that have come down to us : the suit at law brought against him by the Athenians, his exile in Elis, during which he executed works for Olympia, and his death in the first year of the eighty-seventh Olympiad (432 B.C.), are well known. Few names were more popular than his in the ancient world ; legend gathered about it, and in the Middle Ages, when there remained but a con-fused remembrance of classical antiquity, the name of Pheidias survived oblivion. A Roman chronicle of the twelfth century A.D. represents him as a famous philosopher, who visited Rome at the time of Tiberius. The Romans of the Middle Ages found traces of this fabulous journey in the marble horses which are to-day seen on Monte Cavallo; on one of the pedestals is an inscription dating from the Renaissance (OPVS PHIDIÆ), which consecrates the popular legend.

We can form an idea of the masterpieces of Pheidias only by the aid of ancient writers and of a few monuments still preserved.

The Athena Parthenus, placed in the Parthenon in the third year of the eighty-fifth Olympiad (438 B.C.), in the archonship of Theodorus, has been minutely described by Pausanias :—" The statue of Athena is made of gold and of ivory. On the top of her helmet is a sphinx, and on either side are griffins . . . The statue is a standing figure, robed in a long tunic; upon her breast she carries the head of Medusa in ivory. The Victory which she holds in one hand is about four cubits high. In the other hand the goddess holds a spear; at her feet is her shield, and near the spear a serpent, said to symbolise Erichthonius. Upon the pedestal of the statue is represented the birth of Pandora."* The shield was decorated on the exterior by sculptures, portraying the battle of the Amazons, and on the interior was represented the war of the gods and giants.

We cannot dwell upon the different attempts made to restore the Athena Parthenus—either those made on paper, like that of Quatremère de Quincy; or those made in sculpture, like the Minerva of the sculptor Simart, executed for the Duc de Luynes. Materials for restorations are to be found in some

---

* Pausanias, i. 24. 5—7. Cf. Pliny, N.H. xxxvi. 18—19.

ancient monuments, in which clear imitations of the

FIG. 49.—THE LENORMANT PALLAS.
(Athens.)

work of Pheidias may be recognised; the details

of the head and of the helmet seem to be accurately
enough reproduced in a gem cut by Aspasius, now in
the museum of Vienna, and on Athenian tetradrachms
of a date subsequent to Pericles.   The attitude of the
statue is given on a statuette preserved in Athens (the
Lenormant Pallas), which seemed to Charles Lenormant to be a careless reduction of this masterpiece of
Pheidias.*   In this statuette the goddess wears the
ampechonion, a short tunic worn above the long tunic,
and a heavy ægis on her breast. (Fig. 49.)

Very lately, in 1881, was discovered in Athens,
near the Varvakeion, a new statue of Athena, a
product of the second century A.D., which seems
to be an imitation of Pheidias' work.† This is not
a mere sketch, but a marble more carefully executed.
The helmet is decorated with the sphinx and griffin
described by Pausanias ; the ægis, with a fringe of
serpents, covers the chest, and the costume is identic-
ally the same as that of the Lenormant figure.   But
the new marble shows us some details entirely individual : thus the right hand of the goddess holds
a Victory set obliquely and turned towards the spectator; in order to sustain this weight, the hand rests
upon a small column placed in one corner of the
pedestal.   Undoubtedly, however, this statue offers us
the most trustworthy materials known for the restoration of the Athena Parthenus.

---

* A. Conze, *Die Athenasstatue des Phidias im Parthenon*, 1865.

† *Bull. de Corr. hellénique,* January, 1881 ; *Mittheilungen des
deutschen Archäol. Inst.,* 1881, plates I., II.; Th. Schreiber, *Die Athena
Parthenos des Pheidias,* 1883.

There might here be mentioned a long series of monuments which had their origin, more or less directly, in this statue. But though we may by means of these imitations perceive the attitude and costume of the goddess, it is difficult to bring before the imagination the aspect she must have presented with the flesh treated in ivory, the eyes of precious stones, and the drapery executed in gold of different tints by the stainers of gold (βαφεῖς χρυσοῦ) who worked under Pheidias.

The Zeus, consecrated in the temple at Olympia, with the inscription, " Pheidias, son of Charmides, made me," aroused throughout Greece universal admiration ; to die without seeing it was regarded as a misfortune. The description of Pausanias shows it to us seated on a throne of gold, ivory, marble, and ebony, which was decorated with figures both in high-relief and in low-relief;* on the back of the throne were the Seasons and the Graces (Charites), and about the base the great divinities who make up the sovereign council of the gods. The seated god was in a calm and impressive attitude, holding in one hand a Winged Victory and in the other his sceptre ; he was clad in a mantle that left bare one shoulder and the chest, while it covered the lower part of his body with its golden folds, which were enamelled with flowers ; his feet rested upon a footstool adorned with golden lions and figures representing the contest between Theseus and the Amazons. This is the

* Pausanias, v. 11. 1—9.

attitude of the god on a coin of Elis, which gives an authentic reduction of the statue. The type of the face, according to the well-known anecdote, was suggested to Pheidias by two verses of Homer (*Iliad*, i. 528—530).* It is clear, in the first place, that Pheidias imparted to the face of the god a serene and solemn beauty that is reproduced in several ancient busts, imitations more or less faithful of the Olympian Zeus ; it is sufficient to mention the Otricoli bust in the Vatican, and the Zeus Verospi of the same museum. But the most perfect copy is that furnished on a beautiful coin of Elis : the hair, falling simply upon the neck, is bound by a wreath of olive ; the expression of the face is one of calm majesty, and is full of sweetness.†

The Athena Parthenus and the Olympian Zeus, from the point of view of technique, belong to the class of statues known as *chryselephantine*, because their materials are principally gold and ivory. This species of statuary sprang directly from the painted, or *polychrome* sculpture, which flourished vigorously at the time of Pericles, but it was the richer variety. Greek sculptors knew how to obtain marvellous effects from gold and ivory ; the warm tones of the ivory employed for the naked parts of the body give an appearance of life to the statue, and render as well as could be desired the tints of the flesh ; the gold, now red, now brown or greenish, offered

* Strabo, viii. p. 353; Macrobius, *Saturn.* v. 13.

† *Cf.* the restoration attempted by Quatremère de Quincy, *Le Jupiter Olympien*, 1814; compare also L. de Ronchaud, *Phidias, sa Vie et ses Ouvrages*, 1861.

surprising resources for colour effects. Modern taste finds difficulty in admitting the idea of polychromatic sculpture ; the Renaissance has accustomed us to seeing statues clad only in the cold and uniform colour of marble and of stone ; we can with difficulty conceive of that combination of the sculptured form with colour admired by the Greeks. We cannot, however, refuse to acknowledge a form of art which has produced masterpieces ; we must not apply modern theories, which strictly separate sculpture and painting, to Greek plastic art ; our prejudices must give way before duly established facts.

§ 3. THE MARBLES OF THE PARTHENON.

Though we are reduced to conjectures with reference to the masterpieces of Pheidias, the Parthenon marbles furnish us with invaluable testimony for the due appreciation of at least a part of the work conceived by him, if not executed by his own hand. Through divers catastrophes these marbles have at last come down to us.* Early converted by two obscure Byzantine architects into a church, under the name of the Mother of God ($\Theta\epsilon o\tau \acute{o}\kappa o\varsigma$), the Parthenon became a mosque after the capture of Athens by Mahomet II. (1456) ; its history is then forgotten until the point when the Greek Zygomala ascribes its statues to Praxiteles. Up to this time the marbles

* Cf. the work of Michaëlis, Der Parthenon, 1870—71 ; De Laborde, Le Parthenon, 1848 (unfinished); Beulé, L'Acropole d'Athènes ; Petersen Die Kunst des Pheidias am Parthenon und zu Olympia, 1873 ; Waldstein, Essays on the Art of Pheidias, 1885.

L

were almost intact, if we are to judge from the sketches of San Gallo (1465). In 1674, at the time when De Nointel was French ambassador at Constantinople, Carrey, a draughtsman, under the directions of De Nointel, drew the statues of the pediments, the metopes, and the frieze. These drawings are of the greatest value in restoring the sculptural decoration of the temple. The Parthenon was in existence almost as a whole when the Venetian army of Morosini and of Königsmarck laid siege to the Acropolis (1687) ; a bombshell, aimed by a Lüneburg lieutenant, burst through the roof, and made a large breach in the centre of the temple. Entering within the Acropolis, the Venetians destroyed a portion of the statues. Finally, in the earlier years of the present century, Lord Elgin completed the spoliation of the Acropolis by carrying away more than two hundred feet of the frieze, and almost all that remained of the pediment sculpture. These spoils enrich the British Museum.

The only information left us from antiquity as to the two pediments is furnished in a sentence of Pausanias :—" The subject of the front pediment is the birth of Athena ; that of the opposite pediment is the struggle between Poseidon and Athena for the possession of Attica."* We are not able to group the fragments into the original design, except by the aid of Carrey's drawings. Of the eastern pediment, representing the birth of Athena, there remain nine

* Pausanias, i. 24. 5.

fragments, now in the British Museum, and one other still in its place in the Parthenon. Arranging them in the following order, passing from left to right, we are able to restore the scene as conceived by the Athenian master. First appears the Titan Hyperion, guiding his horses as they emerge from the waters ; then a seated figure known as Theseus (or Dionysus),* admirable in style, of energetic design and finished execution ; further along come Demeter and Core together ; behind them Iris runs to announce to the world the birth of the goddess. The centre of the scene is lacking, and can be supplied only upon conjecture.† On the right, a male torso, a fragment of a Victory, with expanded wings, the wonderful group called the Fates, where Pandrosus is doubtless to be recognised, and two of the Seasons (᾽Ωραι), Thallo and Carpo. Finally, the chariot of Selene, descending into the waters, closes the composition, the several parts of which were arranged after the laws of studied symmetry. The groups of statues in a half-reclining position correspond to each other with perfect harmony, as do the various parts of strophe and of antistrophe in the ancient chorus; while the figures of Day (Hyperion) and Night (Selene), closing in the scene, seem to show that it had for its theatre the sky, glowing with sunlight.

The western pediment is more mutilated. In the principal fragments are recognised the seated figure of

---

* [Brunn sees Mount Olympus, scene of Athena's birth, in this figure : *Sitz.-Berichte d. Bayer. Akad.*, 1874.]

† Benndorf, *La Nascita di Minerva, Annali dell' Inst.*, 1865.

L 2

a river-god, doubtless the Cephissus; a group composed of Aglaurus and Cecrops; a part of the body of Athena; a powerful torso, which can be that of no

FIG. 50.—DEMETER AND CORE.
(Eastern Pediment of the Parthenon.)

other than Poseidon, "with mighty chest." Carrey's drawing shows that the artist chose for his subject the instant when the two divinities are in each other's presence. By a blow of his trident Poseidon

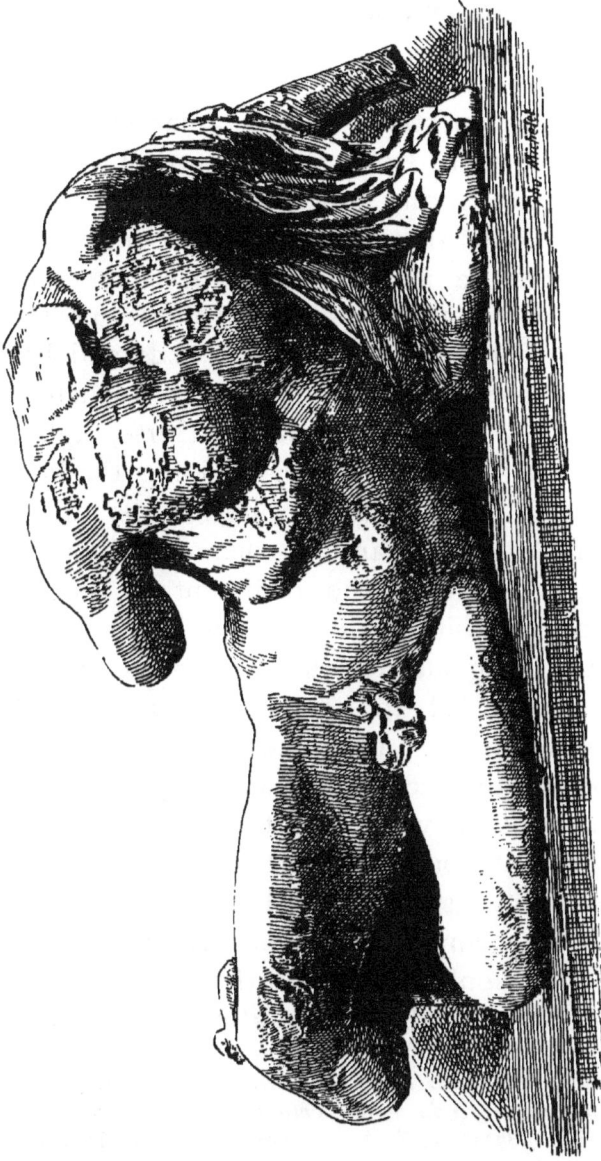

FIG. 51.—THE CEPHISSUS.
(Western Pediment of the Parthenon.)

has just made a spring of water gush forth—a sign of his power. He starts back in amazement before Athena, who, with lance still poised, has caused an olive-tree to spring forth from the rock of the Acropolis. Some archæologists have found the same scene depicted on a vase from Kertch.* These two divinities occupy the centre of the pediment, and behind them are grouped the gods and heroes commonly associated with them : near Athena, are the divinities of Attica, Pandrosus, Herse, Aglaurus, and their father Cecrops, Victory guiding the horses ; near Poseidon, are Thetis, Amphitrite, the divinities of the sea, Aphrodite, and, further on, Ilissus, who occupies this angle of the pediment as Cephissus occupies the opposite angle. The aid of colour was employed to bring out the figures with due prominence. The background of the tympanum against which they stood was painted blue, and was bordered with a red moulding ; the accessories were of gilded bronze.

The metopes have shared the fate of the pedimental statuary ; but little of them is left. Of the ninety-two metopes in high-relief that adorned the portico of the temple, there remained after the explosion in 1687 only thirteen on the north and seventeen on the south;† those decorating the eastern and

* De Witte, *Mon. grecs de l'Association des Études grecques*, 1875. *Cf*. Stephani, *Compte rendu de la Commission arch. de Saint Petersbourg pour* 1872 (1875), and E. A. Gardner, *Athene in the West Pediment of the Parthenon* (*Journ. of Hellen. Studies*, Vol. III.), 1883.

† A single metope remains in place ; fifteen are in the British Museum, one in the Louvre, and one in the Museum on the Acropolis in Athens.

western façades were then in place, but were afterwards broken into fragments by the Turks. In attempting to find unity in the subjects portrayed, we are obliged to resort to conjecture at many points. On the east, the scenes in each of the metopes were taken from the War of the Gods and the Giants, where Athena was figured at the side of Zeus. From other sources we are informed that the young maidens of Athens, devoted to the service of the goddess, embroidered upon her peplus her exploits in this conflict; such a subject, therefore, easily attaches itself to the religious tradition of Athena. On the west, the metopes alternately represent a struggle between a foot-soldier and a mounted warrior, and between two persons on foot. Michaëlis recognises here the conflict between the Amazons and the Athenians,* also represented on the shield of the goddess. On the north, the extremely mutilated condition of the metopes renders great caution necessary in their interpretation ; on a very plausible hypothesis they represent scenes from the Trojan war. The metopes on the south are in a better state of preservation, and in them may be easily recognised the conflict between the Lapithæ and the Centaurs, with scenes taken from the Attic myths, such as the myth of Demeter and Triptolemus, of Pandora and Epimetheus, of Aglaurus and Herse, daughters of Cecrops, who threw themselves from the heights of the Acropolis for having violated the secret of Athena.

In the conflict between the Centaurs and Lapithæ

---

* Michaëlis, *Der Parthenon*, p. 248

we can happily appreciate the success with which the artist overcame the difficulties in a somewhat monotonous subject. In each of the metopes, which invariably represent a Centaur fighting with a Greek, the conflict is expressed with different feeling : here a Centaur leaps upon the body of his foe with all the pride of victory; there another pauses, as if struck with pity, before the half prostrate form of a young Greek. The artist has not shrunk from most realistic details, which show a direct imitation of nature. The workmanship of the metopes is unequal ; but we may still believe that they were executed by artists grouped about Pheidias. The marbles are coloured in part ; the background of one of the metopes found at the Parthenon was red, and the draperies green. Paccard has likewise noticed other traces of red, but these indications are not sufficient to enable us to restore the whole in colour.

Most of the fragments of the frieze which was around the cella are in London ; the museum on the Acropolis possesses a few, and the western frieze is still in place. We know that the whole of the frieze represented, in an unbroken series of subjects, the ceremonies of the Panathenaic festival. The eastern frieze, above the temple entrance, shows the sacred rites performed in honour of Athena Polias by the maidens known as Arrhephori, and by the chief priestess. This central subject is set between groups of the gods that have their sanctuaries near the Acropolis—on the one side Æsclepius and Hygieia, Poseidon, Aglaurus, and Pandrosus ; on the other, Zeus, Hera, Ares, and others, who in majestic attitudes seem to watch the procession passing in the distance. The plan of the composition is simple though grand: the procession

FIG. 52.—COMBAT OF A GREEK AND A CENTAUR.
(Metope of the Parthenon.)

advances in two detachments, divided in order to pass along each of the long sides of the temple, though united at the starting point in the western façade. After this plan, at once symmetrical and

harmonious, advance the old men of the Attic tribes,
resting upon long staves; young girls, clad in robes
that hang in straight folds, carrying pateræ and vases;
the daughters of resident aliens (μέτοικοι), carrying
chairs and parasols destined for the young Athenian
women of free birth.   Then follow the sacrificial
victims, oxen and sheep, the gift of Athenian colonies, guided by young men; then the sons of resident aliens bearing trays and amphoræ; flute-
players and players upon the cithara; also the
thallophori, old men with branches of olive in their
hands; finally war-chariots mounted by the *apobatæ*
and their charioteers in long tunics, and a cavalcade
of horsemen galloping onward at a varying pace.
The frieze on the western front shows the preparations of young Athenians about to join the procession, some of whom are already mounted, while others
are standing near their horses.

Undoubtedly the hand of Pheidias is not to be
discerned in the execution of the frieze.   Portions of
it betray the harsh style of the old Attic school.   The
masters who wrought under Pheidias had not been able
completely to free themselves from their earlier traditions.   But the composition is so grand and so free in
its design, so thoroughly in harmony with the remaining decorative sculptures of the edifice, that we must
believe that Pheidias made the design for the frieze,
though he did not execute it in person.   In its totality
the frieze is wonderfully characteristic of the style of
the school of Pheidias, as it prevailed long after the
masters death.   This style, noble and flowing, the

FIG. 53.—FIGURES FROM THE PANATHENAIC PROCESSION.

(Parthenon Frieze.)

flower of beauty, is the most perfect expression of the
genius of Greece at its most brilliant epoch. After
long toil Greek art had evolved its finest qualities—
simplicity and exquisite and sober taste—which sought
the harmony of the whole before all else. We often
speak of the ideal of Greek art, but we must always
remember that Greek art, even at its best period,
never ceased to draw inspiration from nature. If we
examine the frieze in detail, we shall find that the
portion of it due to artistic conventionality is very
small; even in attitudes and costumes nothing is
artificial. .The artist, with startling fidelity, has
rendered details taken from life itself; the ideal is
nothing more than beauty made real to the sight, but
this is ennobled by a peculiar charm that baffles
analysis, a charm which only long acquaintance
with ancient marbles enables one to feel in all its
delicacy.

In spite of differences in execution, the marbles of
the Parthenon have a certain character in common,
due to the influence of Pheidias. But what is exactly
that part in which we may recognise the actual hand
of the master? The question is a difficult one to
answer. The execution of the figures of the eastern
pediment is generally attributed to him, while those
of the western pediment are regarded as the work of
one of his pupils. We may thus form an accurate
conception as to the Pheidian style, which sums up
and unites in itself the progress of all the Greek
schools in the fourth century B.C. Pheidias was not
only an Attic Greek; trained among Dorians, he

represents the genius of Greece in its general type.
If the Seasons and the group of Demeter and Core
give evidence of the qualities of purest Attic art,
the Heracles and the Ilissus show to what a degree
Pheidias had made his own the energy and strength
of the Dorian style. This is a unique epoch in the
history of Greek art, an epoch when an Athenian
school, by the achievements of one of its masters,
personifies, as it were, the genius of Greek art, with all
its varied qualities.

§ 4. THE ATTIC TRADITION IN THE FIFTH CENTURY B.C.

The genuine Attic tradition, however, was not
lost. It is found again, with its taste for finish and
elegance, in Alcamenes, who, though born in Lemnos,
was Athenian in style. The dates that include the
period of Alcamenes' productivity are between 438
B.C. and 431 B.C., the date of the pediment of Olympia,
on the one hand ; and on the other, 403 B.C. or 402
B.C., the date of the statues executed by him for the
temple of Heracles in Thebes. He is thus a contemporary, and, according to some ancient authorities,
a rival, of Pheidias. Among all his statues adorning
the most celebrated temples of Athens, the one most
admired was the Aphrodite of the Gardens (ἐν
κήποις), of which Lucian describes certain features
as models of elegance and of exquisite art.*
At the time of Dionysius of Halicarnassus it

* Lucian, *Imag.*, 4 and 6.

required a certain amount of study to distinguish between the works of Alcamenes and those of Pheidias.*

The monuments in which the Attic tradition, expanded by the influence of Pheidias, is clearly evident are named below. When work on the Erechtheum, suspended during the greater part of the Peloponnesian war, was resumed in the fourth year of the ninety-second Olympiad (409 B.C.— 408 B.C.), a portion of the sculptures had already been executed. An inscription, giving a list of the works then completed, furnishes us with detailed information as to the sculptures of the frieze, with the names of the artists and the cost of the pieces already delivered.† There are still preserved a few fragments of the frieze, the subject of which was undoubtedly taken from the myth of Erechtheus and the Cecropidæ. The work upon them is very fine, and the effect of it is much heightened by the sombre tint of the Eleusinian marble, upon which, as a background, are placed figures in the white marble of Paros. The part of the temple known as the Pandroseum is adorned with a sort of portico raised upon an upper pedestal-like floor, and made up of an entablature supported by statues of young maidens (αἱ κόραι, or Caryatides). They wear the Attic costume, the *hemidiploïdion*, the tunic, and a small peplus; their hair, gathered into tresses, supports

* Dionysius Halic., *De admir. vi dicendi in Demosthene*, 50, p. 1,108 (Reiske).
† *Corpus Inscr. Attic.* I., No. 324.

FIG. 54.—A VICTORY UNTYING HER SANDAL.
(Fragment of the Balustrade of the Temple of Nike Apteros.)

the spherical capital upon which rests the entablature.

The sculptures of the temple of Nike Apteros
("Wingless Victory") are not all of the same date.
The frieze,* the subjects of which were suggested by
the victories of the Athenians over either barbarians
or Greeks, resembles in style that of the sculptures on
the Theseum ; it is without doubt of the same date
as that temple, and anterior to the Propylæa and
the Parthenon.   But around the little temple extends
a balustrade, adorned with bas-reliefs exquisite in
style.†   These sculptures, the principal fragments of
which were discovered by Hansen and Schaubert, are
later than the rest of the temple, and perhaps date
from the administration of Lycurgus ; but nothing
can better illustrate than do these sculptures the style
of the Attic school at the end of this period, which
borders upon that of Praxiteles.   The figures of
Victories, messengers of Athena, symbolise the
triumphs of the Athenian people : one leads a bullock
destined for sacrifice ; another removes her sandals as
if she had hastened from some field of battle to
announce the success of Athenian arms.   Their bodies,
very finely modelled, are outlined under light draperies
with most delicate folds.   The style, less grand than
that of the Parthenon, clearly shows a return to the
distinctively Attic tradition.

* The eastern and southern portions are now in place ; the other
portions are in the British Museum.

† On these bas-reliefs compare R. Kékulé, *Die Reliefs an der
Balustrade der Athena Nike*, 1881.

We now return to the age of Pheidias, in order to
follow the development of sculpture in Peloponnesus
to the middle of the fifth century B.C. Recent discoveries have completely reshaped its history, and
the interesting marbles of Olympia have revealed to
us a school hitherto almost unknown.

In the valley of the Alpheus, at the point where
the Cladeus unites with it, rose at the foot of Mount
Cronius the Doric Temple of Olympian Zeus, begun
in 480 B.C. by Libon, of Elis, and finished not more
than about thirty years later. Near at hand were
the temple of Hera, the Metroön, and the Treasuries,
built by various Greek cities, all combining to make
Olympia a holy city ; while encompassing the temple
of Zeus was the sacred grove or Altis, with countless
statues consecrated within its enclosure by the piety
of the Greeks. It had long been known that the pagan
emperors had drawn largely from this treasury of works
of art, and that the Christian emperors, Constantine
and Theodosius, had robbed Olympia for the sake
of Constantinople. It was hoped that the sculptural
decorations of the temple might be uncovered from
beneath the alluvial deposits of the Alpheus, which
had gradually accumulated with the lapse of time.
In 1831, the French expedition to the Morea began
excavations which, though too soon broken off,
brought part of the temple of Zeus to light.* The

* Dubois and Abel Blouet directed the excavations.

M

work was again taken up by the German Government, and the seasons of excavations that have followed since 1875 have not ceased to be fruitful.* In the museum at Olympia, at the foot of Mount Cronius, we to-day possess a noteworthy part of the marbles that adorned the temple ; with them are preserved many fragments of all kinds, and many inscriptions.

The simple, though definite, descriptions of Pausanias make known to us the subjects represented on different parts of the edifice. The metopes, twelve in number, portrayed the labours of Heracles, and were placed under the colonnade above the doors of the pronaos and of the opisthodomos. Two magnificent fragments of the metopes were discovered by the French expedition to the Morea, and are now in the Louvre.† The most beautiful, that of Heracles conquering the Cretan bull, is a powerful composition ; the hair and the beard are treated in masses; the modelling of the flesh is firm, and the lines of the group are bold. The second, where Athena, seated, looks upon the struggle of the hero with the Stymphalian birds, shows the same characteristics, though in a lesser degree. The German excavations led to the discovery of a large number of fragments from the metopes, which have supplemented the metopes

---

* The excavations have been conducted, under the chief direction of E. Curtius, by Bötticher, Adler, Hirschfeld, and G. Treu. Compare *Ausgrabungen aus Olympia*, Berlin, 1879—1881, with photographic eproductions ; also Bötticher, *Olympia*, 1883 ; and Overbeck, *Griechische Plastik*, 3rd ed., 1881—82.

† The excavations of Dubois and Blouet also brought to light some other fragments, the most important of which is the Nemean lion.

of the Louvre, showing us in part at least other

FIG. 55.—HERACLES AND THE CRETAN BULL.

scenes from the labours of Heracles: the conflict of
the hero with the Nemean lion, the Lernean Hydra,

M 2

Geryon, the Erymanthian boar, etc.*   That which
we reproduce (Fig. 56) shows Heracles, aided by one
of the Hesperidæ, sustaining the world upon his
shoulders, while Atlas presents to him the golden
apples of the Hesperidæ. Though with very realistic
qualities, the style of the metopes is not altogether
free from the conventionalism of archaic sculpture ;
the attitudes of the figures and the execution of the
draperies indicate a date anterior to that of the
metopes of the Parthenon. It is very probable that
these are the work of Peloponnesian artists, employed for the decoration of the temple ; for, forming part of the edifice, these metopes must have been
executed at the same time with it, before the arrival
of the Athenian masters who laboured at Olympia.
It is by no means a mere hypothesis to place their
date a little before the sculptures of the Theseum.

The metopes are the work of unknown masters,
but tradition preserved at Olympia the names of the
artists of the pedimental sculptures. Pausanias took
pains to note that the eastern pediment was the work
of Pæonius, born in Thrace, at Mende on the Hebrus,
a colony from Asiatic Ionia. Remaining in Olympia
after the death of Pheidias, he was entrusted, after
competitory trial, with the execution of the acroteria
of the temple, i.e., the gilded Nike which surmounted

---

* We shall be in a better position to judge of the importance of
these discoveries when the work of reconstructing the metopes, under
the direction of Treu, shall have been completed. Compare the
pamphlet which serves as a guide to the casts from Olympia in Berlin,
Die Abgüsse der in Olympia ausgegrabenen Bildwerke, 1880.

it and the two great vases at its extremities (between
430 B.C. and 422 B.C.). He was still living at Olympia,

FIG. 56.— HERACLES, ATLAS, AND ONE OF
THE HESPERIDÆ.
(Metope of Olympia.)

when the Messenians, after the victory at Sphacteria,
ordered of him a statue of Nike, which, with an

inscription at its base, has been found in the excavations.*

The body is thrown forward, outlined beneath
folds of drapery blown by the wind. It is the
work of a master full of fire, inspired by living reality,
who has entirely escaped from the influence of the
schools. His qualities are distinctly the opposite of
those of the Attic school. Pæonius appears as the
freest and most individual representative of Peloponnesian art ; he reveals the astonishing freedom of
Greek genius at its epoch of perfection ; at the same
time he acquaints us the better with the Peloponnesian masters, of whom Pheidias was a pupil. He
explains to us one side of the genius of Pheidias,
that which passed beyond the somewhat narrowing
limitations of Attic tradition.

The eastern pediment represents Pelops making
ready to contend with Œnomaus in the chariot race
which the king of Pisa required of all the suitors of his
daughter, and in which Pelops was the victor. The two
contestants and their attendants form two groups, between which stands Zeus, judge of the contest, an erect
figure occupying the centre of the pediment. On the
right Hippodameia and Pelops, Sphærus, the esquire
of Pelops, with four horses, two servants seated, and
in the corner the river Alpheus reclining ; on the left,
Œnomaus, his wife Sterope, his charioteer Myrtilus,

---

* "The Messenians and Naupactians have consecrated this statue to
Olympian Zeus, as a tenth of the booty captured from the enemy.
Pæonius of Mende made it, and for the acroteria placed upon the
temple he won the prize." [Cf. Hicks, Greek Hist. Insc., 1882, p. 81.]

his horses, a servant, a young maiden, and the figure
of the river Cladeus. We find here again those conditions of absolute parallelism which assert themselves
in the composition of the pediment, and the seated
or reclining attitudes, in which the several figures are
placed, as required by the triangular shape of the
pediment. The principal figures rescued by the excavations are much mutilated. There is a concurrence
in recognising in some superb fragments the powerful
torso of Zeus, the standing figure of Pelops, with
a haughty bearing, and Hippodameia clad in an
ample Dorian peplus, with heavy straight folds.
The subordinate figures are better preserved; one
of the servants of Pelops in a crouching position,
and the kneeling charioteer of Pelops, are almost
uninjured, and have attitudes at once natural and
plastic. The figures in the corners may also be
recognised, the Alpheus and the Cladeus. The
former, resting upon his elbow, seems to watch the
preparations for the contest; the modelling of the
body is gentle, its outlines are not strongly marked.
In the execution of the Cladeus, however, the figure
is energetically rendered, and in the swell of the
tense muscles and in the firmness of its contours
is to be recognised the chisel of a powerful master.
We give (Fig. 57) the head of an old man who is
seated, without doubt one of the servants of Pelops.
His head is held erect, and he looks forward. There
are few Greek monuments where the feeling of actual
life shines forth more intensely than here; details
taken from life are not disregarded—the baldness, the

wrinkled forehead, the humble and submissive expression. The details of all these parts taken together
are of rapid and vigorous execution, with occasional
striking evidences of carelessness. Thus, the backs
of several statues are entirely unfinished ; it seems as
if work was carried on in great haste, under the stress
of time. It is a question whether the style of these
marbles permits us actually to attribute the eastern
pediment to Pæonius. It is difficult not to note the
profound difference between these marbles and the
Nike, an authentic work of the master. The art of
the eastern pediment is rude, and inferior to that of
the Nike. In order to bring into accord the testimony
of Pausanias with that of the marbles, we must choose
between several solutions, the most probable of which
seems that the pediments were executed according to
the designs of Pæonius by artists of Elis whose talents
were very unequal.

The same problem presents itself in the western
pediment, which Olympic tradition, preserved by
ancient writers, assigns formally to Alcamenes. The
subject is the battle of the Centaurs and Lapithæ at
the wedding of Peirithoüs, a subject frequently treated.
We are aware how few were the innovations as to
subjects made by the masters of the best epoch.
Pausanias has described this pediment less completely
than the eastern pediment. In the centre is Peirithoüs ;
at the left is Eurytion, who has just seized Deidameia,
and "Cenæus bringing aid to Peirithoüs ; on the other
side Theseus is striking Centaurs with his axe ; among
the Centaurs, one is carrying away a young girl,

another a boy of rare beauty."* There were twenty-
one figures in the scene; the excavations have brought
to light fragments of all these figures, which have

FIG. 57.—HEAD OF SEATED OLD MAN.
(Eastern Pediment at Olympia.)

been skilfully restored by Treu. The central group
of the whole composition is made up of seven persons:
in the middle of the pediment, standing erect, is
Apollo, who with arms extended dominates the

* Pausanias, v. 10, 8.

whole; his torso is of rare elegance, and the head, with hair curled in the archaic style, is charming in its youthful expression. On the left comes first the Centaur Eurytion seizing Deidameia, who defends herself against her ravisher. Peirithoüs, following next, is rescuing his bride. To the right of Apollo, from the spectator's point of view, a Centaur has seized a young maiden with a brutal movement, who, her charming face preserving a calm expression in the midst of her danger, is struggling in his arms. Theseus, of whom only fragments remain, is swinging his axe to save the young Greek maiden. On either side of the central group the subordinate groups mutually correspond in accordance with the laws of symmetry. At the spectator's left is one of the young Lapithæ (Fig. 58) struggling with a Centaur; then, a woman kneeling, whose hair is clutched by a Centaur; a Greek, who seems to be rising in order to join in the fray; an old woman greatly terrified, who looks upon the scene with horror; and, finally, a nymph extended in the corner of the pediment. In the portion to the right is placed a group consisting of a Centaur carrying off a young boy, a woman in a Centaur's clutch, a Greek warrior armed with a sword, and, finally, under the lowest part of the coping, a waiting-woman and a nymph, who correspond to the figures in the other corner.

Nothing can be more skilful than the composition of this pediment; the whole space is occupied with works of finished art, and the figures have a life and movement that give to the whole a dramatic effect.

The style of these marbles is not less surprising than

FIG 58.—ONE OF THE LAPITHÆ.
Western Pediment of Olympia.)

that of the eastern pediment, to which it bears a

remarkable resemblance; the same rapid and careless work, the same inaccuracies, which are inexplic-
able if the hand of Alcamenes sculptured these figures.
How can it be admitted that a rival of Pheidias could
have modelled these feeble draperies, under which it
is impossible to perceive the forms of the body, or
these Centaurs, with their extraordinary anatomy?
It must have been that the master merely designed
the scene, and left the execution of it to the same
artists who sculptured the eastern pediment. This
supposition explains at least two incontestable facts—
the resemblance in style between the two pediments,
and the inequality of workmanship, as contrasted with
the skilful art in the compositions.

The Marbles of Phigalia.—Attic influence is once
more apparent in the marbles found at Phigalia. These
marbles form the frieze of the temple of Apollo Epicurius, erected in 430 B.C. (Olympiad LXXXVII. 3) by
Ictinus, architect of the Parthenon. Pointed out in 1765
by Bôcher, a Frenchman, the temple was excavated in
1812 by the archæologists who had discovered the marbles of Ægina, and the sculptures, bought by the Prince
Regent of England, now enrich the British Museum.
The frieze represents a double subject: the conflict
between the Lapithæ and the Centaurs, and the war
of the Athenians and the Amazons. It is a matter
of doubt whether this work is from the hand of an
Athenian artist, introduced by Ictinus, or whether it
was executed on the spot by a local sculptor. The
most probable conjecture is that the composition
alone is due to an Athenian, to which theory the

mediocre execution gives additional strength. In its style the composition attempts by a certain nicety of execution to give freshness to a subject already somewhat worn out. Skilful as is the arrangement of some of the figures, there are traces of false taste in the details; the violent and contorted attitudes of the Amazons, the singular movements of the Centaurs, who kick and bite at the same time, are more ambitious than dramatically effective. We seem to have before our eyes the work of a provincial artist, who wished to improve upon the Athenian masters.

§ 6. SCHOOL OF ARGOS.—POLYCLEITUS.

At another locality in Peloponnesus, the Argive Polycleitus was continuing the traditions of the school of Argos, from which that of Sicyon must not be separated; the Sicyonian sculptors, in fact, studied at Argos, and then returned to their native city. Polycleitus, born about the seventy-fourth Olympiad (482 B.C.—478 B.C.), was but a few years younger than Pheidias, and, like him, was a pupil of the Argive Ageladas. A list of his works that may be drawn up from the ancient writers shows us that he borrowed his subjects from religious and heroic cycles; he also continued the tradition of the Peloponnesian schools in executing statues of athletes for Olympia. The Heræum of Argos possessed one of his chief works, the famous statue of Hera, in which he fixed, as it were, the classic type of that goddess. The figure, made of gold and ivory, was seated upon a throne, and wore a crown, on which were represented the Charites and the Seasons; in one hand the goddess held a pomegranate, and in the other a sceptre, surmounted by a cuckoo. The principal features of the type thus created by Polycleitus are found on the beautiful coins of Argos, where is to be seen the crowned head of Hera, but above all in the magnificent colossal bust of the Villa Ludovisi, at Rome.

At Ephesus was to be seen the statue of an Amazon, the work of Polycleitus. The Ephesians, on comparing it with similar statues made by other masters, gave it the first place; the Amazon of Pheidias came next, then the statues by Cresilas* of Cydonia, and by Phradmon. We know nothing of the Canephori of Polycleitus, nor of the group representing two boys playing with knucklebones (*astragalizontes*), which were transported to Rome. But the beautiful statue of the Villa Farnese, which represents a young man binding a fillet about his head, is undoubtedly a copy of the Diadumenus of Polycleitus, also reproduced in the bronze figured in our text (Fig. 60).† The style is of exquisite refinement; the hair is treated in irregular curls. It is known that at this point the Argive master broke

FIG. 59.— HERA. (From a Coin of Argos.)

* The Amazon of the Lansdowne collection, in Berkeley Square, London, is, perhaps, a copy of the work of Cresilas.

† Numerous replicas of this statue are in existence, among others the Vaison statue in the British Museum.

FIG. 60.—COPY OF THE DIADUMENUS OF POLYCLEITUS.
(Bronze of the Bibliothèque Nationale Paris.)

with the traditions of archaic sculpture, which were still followed by his contemporaries.

Ancient criticism declared that " no one surpassed Polycleitus in refinement of detail."* Perfection in execution was in fact the supreme concern of the Argive-Sicyonian school, and Polycleitus attained it in a rather restricted range of subjects. For him youth was the inseparable condition of beauty, and he seems to have always addressed himself to subjects present-ing youth. " Polycleitus," adds Quintilian, " shrank from the representation of mature age, and never dared go beyond that of beardless youth."† Perfect knowledge of the human body, and perfect accu-racy in detail, were the characteristic features of the genius of Polycleitus. He wrote a treatise on the proportions of the parts of the body, and exemplified his theory in the celebrated statue of the Doryphorus. It is difficult to determine exactly the proportions fixed in his *canon*, which served as the rule to the time of Lysippus, and may be studied in the replicas of Naples, of Florence, and of the Vatican. The description of Lucian is vague enough : " In the body we must conform to the rule of Polycleitus ; it must be neither too tall and long beyond proportion, nor of a stature too short, resembling that of a dwarf, but of medium and appro-priate height."‡ This was something intermediate, a transition from the stunted and massive forms of

---

* Quintilian, *Inst. Or.*, xii. 10.
† Quintilian, *id. ib.*
‡ Lucian, *De Saltatione*, 75.

archaic Doric art to the more slender type caused to prevail by Lysippus. In a study upon the Dory-phorus of Naples, E. Guillaume thus speaks of the meaning of the canon of Polycleitus : "It was the summing-up, as it were, of a past school, and not the point of departure for a new school. At this time Dorian art attains its perfection, both in its buildings and in the sculptural representation of man. Many years later, after having thus realised an ideal in which energy and force are dominant, the Greeks un-dertook consciously to seek an ideal of pure beauty and elegance."*

The influence of Polycleitus upon the schools of Sicyon and of Argos was very great. His pupils remained faithful to his rules, or even went beyond them. It is thus that the Discobolus of Naucydes, the Argive, a copy of which has sometimes been recog-nised in a statue in the Vatican, had undoubtedly the same quietness of attitude that characterised the style of Polycleitus. His other disciples, Alyppus and Polycleitus the younger, are known to us only from the statements of ancient writers.

It has been observed that Greek art, at its best period, offered an infinite variety. It is impossible to sum it up in a sentence, which would necessarily be commonplace and restricted ; the only way to become familiar with it is to study it according to the schools, and thus to ascertain the profound differences that separated one school from another.

---

* *Doryphorus*, in Rayet, *Mon. de l'Art antique*, 3e livraison.

N

CHAPTER V.

FIFTH PERIOD.

TO THE ALEXANDRINE AGE.
FROM THE NINETY-SIXTH TO THE ONE HUNDRED AND TWENTIETH OLYMPIAD (396 B.C.—292 B.C.)

THE fourth century B.C. bears but a slight resemblance to the fifth century B.C. To an age of faith and belief succeeds a period of scepticism; the severe dignity, the seriousness of earlier Greek art, gives place to a more sensuous taste. In the fifth century B.C. art was pre-eminently religious. Though in the creation of divine types in art it drew its inspiration from the living human form, it was only to put reality to the service of a higher conception of beauty; gods were in the likeness of men, but it was the most beautiful likeness possible. The fourth century B.C., however, humanised the types of the divinities; it caused the gods of Olympus to come down to earth and to share the passions of humanity; art became more familiar; it detached itself from religion in order to find in real life an individual and personal character. Already at the close of the preceding period Callimachus had won the epithet of κατατηξίτεχνος from his excessive refinements. Demetrius of Athens carried his realism so far that Lucian names him "the maker of men (ἀνθρωποποιός), and not the maker of statues of men (ἀνδριαντοποιός)." The rhetorician of Samosata describes as follows one of his statues: " Hast thou not seen near the stream a statue with prominent belly, bald head, half nude, with thin beard apparently blown by the wind, and veins standing out as upon a living person ? That is Pelichus, the Corinthian general."*

It was chiefly in the Attic school and in that of Paros that this passion for expressing the intensest and most personal sentiments in art became especially marked. The most brilliant representatives of this school are two masters, almost contemporaries, Scopas and Praxiteles.

§ 1. SCOPAS. †

The two extreme dates that limit the period of the activity of Scopas, who was born at Paros, are the burning of the temple of Athena Alea at Tegea (Olympiad XCVI. 2, 395 B.C.), which Scopas rebuilt, and the death of Mausolus (Olympiad CVII. 2, 351 B.C.), on whose tomb he also worked. He executed for the temple of Athena‡ the two pediments, one of which represented the Calydonian Boar Hunt, the other the combat between Achilles and Telephus on the banks of the Caïcus. Like

* Lucian, *Philopseud.*, 18.
† *Cf.* Urlichs, *Skopas Leben und Werke*, 1863.
‡ Fragments of the decoration of this temple have been discovered in recent excavations at Piali, the ancient Tegea (*Arch. Zeitung*, 1880).

the sculptors of the preceding generation, Scopas counted among his works many statues of gods, an Aphrodite Pandemus in Elis, an Asclepius and a Hygieia at Gortys, with others in Attica, Megara, Thebes, Pergamum, and elsewhere. Some of his statues possess a special interest, because they bear witness to the new conceptions that were asserting themselves in Greek art. Such is the statue of Apollo Musagetes, of which the Vatican possesses a replica. Apollo is now no longer the sturdy Dorian god, so often treated by the archaic masters. Clad in a long robe, the god has an almost feminine bearing, and the sculptors succeeding Scopas have emphasised his characteristics of elegance and delicate art. A Maenad tearing a kid was celebrated for its spirit and for the truthfulness of its attitude. His group of Eros, Pothos, and Himeros, in the temple of Aphrodite at Megara, representing in a plastic form the desires of the soul, shows well from what source the artists of the new school drew their inspiration. These graceful creations owe their existence to the analysis of the emotions of the soul, and sculpture thus seeks more and more for that which shall move and touch the feelings.

This new taste betrays itself in every form of art: this is the time when in architecture the freer and more varied Ionic order supplants the Doric, and makes infinite combinations possible. Scopas assisted in the work on the temple of Ephesus, and adorned with sculptures the shaft of one of the Ionic

These marbles come from the pediments, and are doubtless from the hand of Scopas. *Cf. Mitth. d. deutsch. Inst. in Athen.,* 1881.

columns.* Mr. Wood found at Ephesus fragments of columns decorated in this manner.†

But we have still other means than replicas for studying and appreciating the style of Scopas. The excavations of Mr. Newton at Budrun have exhumed the priceless marbles of the Mausoleum of Halicarnassus, the work of the chisel of Scopas, and of his rivals, which can acquaint us at least with the style of the contemporary Attic school. Artemisia, queen of Caria, widow of Mausolus, had entrusted to several artists the execution of the tomb of her husband : these were Scopas ; Leochares the Athenian, who had already departed from the distinctive features of the Attic school by his taste for portraits ;‡ Bryaxis of Athens; Timotheus and Pythius—the latter, assisted by Satyrus, was also the architect of the Mausoleum. Remaining almost intact until 1402, the Mausoleum repeatedly furnished the materials for a fortress for the Knights of St. John. In 1846 the bas-reliefs built into the castle walls of Budrun were transported to London, and in 1855 Mr. C. T. Newton discovered what was left of the mutilated monument, fragments of the frieze, and of the statues of Artemisia and of

---

* [This statement is based on the received text of Pliny, *N.H.* xxxvi. 95, "columnæ . . cælatæ, una a Scopa." But the true reading may have been, "columnæ . . cælatæ imo scapo " : it is only on the lower part of the shaft (imo scapo) that these columns were sculptured.]

† One of these is now in the British Museum.

‡ He had represented several members of one family : Pasicles and others, as is indicated by pedestals with inscriptions found upon the Acropolis. A copy of his Ganymede Carried away by an Eagle is in the Vatican

Mausolus, parts of the quadriga sculptured by Pythius, of lions, and of architectural members. This débris furnished Pullan and Fergusson with the materials and the data for the restoration of the Mausoleum, a task which each in turn has attempted.

The British Museum possesses the fragments of lions which doubtless formed the decoration between the columns. They are treated with great skill, but in the somewhat conventionalised style which Greek art observes in the type of these animals. The frieze is composed of three distinct parts, one passing along the sides of the edifice, another along the entablature of the portico, and the third upon the substructure. The frieze, of which the most important fragments have been preserved,* is that representing the combat of the Greeks and the Amazons ; of the other friezes, in which the conflict between the Greeks and the Centaurs, and chariot races, are represented, only a few fragments exist. The style is very unequal : it commonly exhibits a certain affected striving after violent movement and difficult attitudes, which makes these friezes resemble those at Phigalia much more than the marbles of the Parthenon. At the same time the proportions of the figures are slender, and the flying draperies give them great vivacity. Scopas certainly put his hand to these marbles ; but we must probably look for the most distinct evidences of his style in the more important pieces, such as the fragments of the group on the substructure and of the two colossal

* The larger part is in the British Museum ; there is one fragment at the Villa Negro in Genoa.

statues, Artemisia and Mausolus. The Carian king is figured as if in an apotheosis, and his head, with thick hair, betrays an interesting attempt at barbaric naturalism.

The figures of the Mausoleum give a remarkable precision to the conception that we may form as to the style of Scopas. It is safe to ascribe to his

FIG. 61.—FRAGMENT OF THE FRIEZE OF THE MAUSOLEUM
AT HALICARNASSUS.
(British Museum.)

school a work which shows a certain kinship with the marbles of Halicarnassus; this is the Victory found in the island of Samothrace, now in the Louvre. It surmounts a votive offering in the form of the prow of a trireme, consecrated by Demetrius as a token of a naval victory won over Ptolemy in the neighbour-hood of the island (306 B.C.). Holding in one hand a trumpet, and in the other a wooden cross, which supported the trophies, the figure of Victory advances proudly with extended wings as if about to soar in flight. If the work is not from the hand of Scopas, it is by one of his pupils ; in it may be discovered that carefulness of expression, that taste for movement and for draperies blown by the wind, which is characteristic of the Parian sculptor.

§ 2. PRAXITELES.

According to Vitruvius, Praxiteles worked with Scopas on the sculptures of the Mausoleum. He thus began the most brilliant period of his career about 357 B.C. ; Pliny, however, gives Olym. CIV. (364 B.C.) as the date when Praxiteles was at his acme. Without discussing this contradictory evidence, we may admit that the principal period of the activity of Praxiteles was between 360 B.C. and 340 B.C. An inscription from Thespiæ, which Boeckh believed referred to the great sculptor, gives Athens as his birthplace ; his sons, Cephisodotus and Timarchus, were Athenian citizens. We know, moreover, that Praxiteles lived in Athens, where his relations with the courtesan Phryne, his model, are well known.

Among the sculptors of the Attic school in the fourth century B.C., Praxiteles best represents the new spirit. The severe gravity of the old Dorian schools is abandoned ; art applies itself to more pleasing subjects, which awaken the inmost sentiments and feelings. This period has unjustly been called a period of decadence ; it is rather an epoch

of development in Greek art, for at no time did
Hellenic genius unfold more brilliantly its exquisite qualities of delicacy. This is the time when
the *coroplastæ* of Tanagra modelled their figurines
with so spirited and life-like an execution ; when the
potters of Athens were decorating, with elegant paintings, the masterpieces of ceramic art ; and when,
in sculpture, Praxiteles and his school conceived their
charming types of goddesses and young satyrs, and
taught sculpture above all to seek grace and delicacy of form.

Not less than forty-six groups or statues executed
by Praxiteles are mentioned. This list cannot be repeated here ; we shall confine ourselves to works of
which copies have been preserved to us. With common
consent, antiquity admitted that Praxiteles had created
the type of Aphrodite, in his celebrated statue for
Cnidus. "Who gave a soul to marble ? Who saw upon
this earth the Cyprian goddess ? Who put into stone the
ardent desire for pleasure ? It was done by Praxiteles.
Olympus is deprived of the Paphian goddess, because
she has descended to Cnidus."* The most authentic
testimony by which we recognise the work of Praxiteles is a medal of Cnidus struck in honour of Plau-
tilla and Caracalla ; it shows the essential features of
the masterpiece, as described by Lucian. The subject
was, furthermore, repeated many times for the cities
of Cos, of Thespiæ, and of Heracleia. After Praxiteles
other sculptors took it up in emulation ; the Venus of

---

* *Anthol. Planud.* iv. 159. See also Lucian, *Imag.* 4 and 6.

the Capitol, the Venus di Medici, and many other replicas in our museums, are derived from the type so happily conceived by Praxiteles. To the cycle of Aphrodite was naturally attached that of Eros ; it was here that the Athenian sculptor must have found the subject of the Eros of Parion, and of the Eros consecrated by Phryne at Thespiæ. ˙ Praxiteles represented this god with all the features of a youth " in the flower of his age," as may be seen in the beautiful torso of Centocelle in the Vatican. The Apollo Sauroctonus is also a new conception. Scopas, in his Apollo Musagetes, represented the inspired god, while the beautiful youth of Praxiteles, with his slender form and delicately-modelled outlines,* has all the grace of an Athenian ephebus. Praxiteles was inspired also by the legends of the cycle of Dionysus, and delighted in treating the figures of the laughing fauns and drunken satyrs that formed the cortège of the god. Perhaps we may recognise a copy of his faun *periboëtos* (" far-famed ") in the charming figure of the Capitoline Museum, where a young satyr, clad in

FIG. 62.—APHRODITE.
(On a Medal of Cnidus).

---

* Copies in the Louvre and in the Vatican.

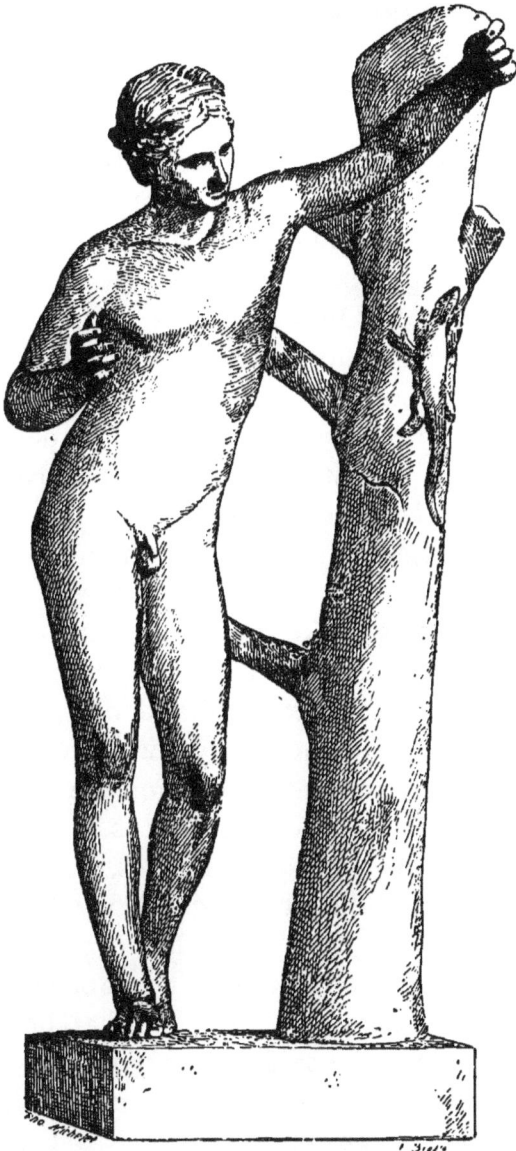

FIG. 63.—APOLLO SAUROCTONUS.
(Museum of the Louvre.)

a fawn skin, leans carelessly against the trunk of a tree.*

We have thus far mentioned only replicas. The excavations at Olympia have brought to light an original work of Praxiteles; it is the Hermes with the infant Dionysus, which was consecrated in the Heræum at Olympia; it is of marble and of exquisite workmanship, and exhibits all the grace of the Athenian master. The god is represented with all the characteristics of a young man of delicate outline, and in his half-smiling face, bent above the tiny Dionysus, who is sitting on his left fore-arm, we find a charm of expression hitherto unknown in Greek sculpture.†

Even from this rapid review we may observe the spirit that inspired Attic sculpture at this time. The representation of athletes is abandoned; among the works of Praxiteles we note very few subjects from real life, such as the warrior and his horse, a group which was to be seen in the outer Ceramicus in Athens. On the other hand, the cycle of goddesses, or of young and charming gods, kindled his inspiration; but whether he represents Eros or Aphrodite, or Demeter and Core, as in the group *Catagusa*, it is, above all, charm of form and delicacy of expression that he attempts to portray; it may be said, in the words of Diodorus Siculus, that he excelled, above all, in rendering " into stone emotions of the soul."

---

* Many other replicas are found in the museums of Rome. [The figure in the Capitoline Museum is Hawthorne's "Marble Faun."]

† Treu, *Hermes mit Dionysos Knaben*, 1878.

Scopas and Praxiteles proceeded from a similar school; even at the time of Pliny it was not known to which of the two should be attributed the statues of the Niobidæ, which were placed in the temple of Apollo Sosianus at Rome, whither they had been brought from Asia Minor, and of which figures the statues in the museum in Florence are copies. Stark, contrary to the view of Welcker,* has proved that this group could not have adorned a pediment, but that the statues must have been placed between the columns of a portico, in such a way that they stood out singly and isolated. Whoever may have been their author, this group belongs to the class of dramatic sculptures much in vogue at that time. In the attitudes of the mother who sees her children pierced by the darts of Apollo, in the frightened or supplicating expressions of the young men and young girls, everything is designed to stir the emotions. It is, as it were, the final scene of a tragedy.

The sculptures of this epoch, known from the Greek and Roman authors and from inscriptions, are very numerous; the further we advance in the history of sculpture, the more do the signatures left by sculptors upon their marbles multiply themselves. A collection of these signatures has lately been edited by

* Stark, *Niobe und die Niobiden*, 1863; Welcker, *Alte Denkmäler*, I., p. 223. Besides the statues in the Uffizi, replicas of isolated figures are in England, Russia (at Tsarskoe-Selo), in the Capitoline and Chiaramonti Museums of Rome, at Aquileja, and at Cologne. Welcker thinks that the group was copied twenty times. According to Overbeck it was composed of thirteen figures including the pedagogue, who holds one of the children.

E. Loewy.* The artists more intimately connected with the school of Praxiteles are Cephisodotus the younger and Timarchus, sons of the great sculptor, who frequently laboured together, and made portraits like those of the orator Lycurgus and of Menander.

The monuments preserved from this period that may be dated with accuracy are not very numerous. It is principally in bas-reliefs accompanied by inscriptions that we can find hints as to dates. Without being monuments of great art, these bas-reliefs exhibit the prevailing qualities of that epoch, and have the advantage of being original works. The most important dated monument among the less interesting works of Athenian sculpture in the fourth century B.C. is the frieze of the choragic monument of Lysicrates, put in place in the earlier years of the one hundred and eleventh Olympiad (335 B.C.—334 B.C.). It represents the defeat of the Tyrrhenian pirates by Dionysus, and their metamorphosis into dolphins. The style, though elegant, is distorted, and very far removed from the calm simplicity that marks bas-reliefs older by a half century, as, for instance, those of the year 410 B.C. and of 375 B.C.,† which were placed as headpieces above some treaties of alliance, and reflect all the qualities of Athenian art at the beginning of ·the fourth century B.C.

Outside Attica we find an important group of

---

* E. Loewy, _Inschriften der griech. Bildhauer_, 1885.

† Museum of the Louvre, and Museum of Athens. Consult, in regard to the latter, _Bull. de Corr. hellén._, II., plate XII., and also _infra_, p. 233.

sculptures which bear a strong resemblance to the marbles of the Mausoleum, and must be referred to the same school; these are the marbles of the tomb of Harpagus, general of Cyrus, discovered at Xanthus, in Asia Minor, by Mr. Fellows. The principal figures, placed in the Lycian room of the British Museum, are four in number, and represent draped women of life size; the marine symbols carved on the basis indicate that they are Nereids. The decoration of the monument comprises also some smaller figures, and a frieze where Ionian hoplites and Lycians in their national costumes are grouped as if engaged in combat. The similarity in composition to that of the marbles of the Mausoleum leads us to believe that these sculptures proceed from a school in Greece proper.

To what period and to what school are we to assign the famous Venus de Milo? While it is universally agreed that in this marble of the Louvre we recognise a masterpiece of Greek art, few monuments have given rise to more animated controversy. Several archæologists have seen in it a copy of a work of Alcamenes; others, with greater probability, have ascribed it to the school of Scopas.* The fourth century B.C. is a time that agrees well with

---

* The opinion according to which the Venus de Milo is a work of Roman times is hardly tenable. It is supported only by one fact : there was found at the same time with the statue an inscription of the first century B.C., with the signature of a sculptor from Antioch on the Mæander : ['Αγή] σανδρος Μηνίδου ['Αντ] ιοχεὺς ἀπὸ Μαιάνδρου ἐποίησεν. Besides, it has not been proved that this inscription belonged to the statue. Cf. Fröhner, Notice de la Sculpture antique du Louvre (1869), 1874, p. 168 ff.

the style of this statue, which is at once severe and charming, and is impressed with an original and thoroughly individual grace. We can hardly class the Venus de Milo among the more or less successful replicas that were multiplied in imitation of the Aphrodite of Praxiteles. The restoration of this mutilated statue has often been attempted ; perhaps the goddess held an apple in one hand, while with the other she supported the drapery which covered the lower part of her body. Ravaisson thinks that it made one of a group of Mars and Venus, and a long series of similar monuments adds force to his opinion.*

§ 3. THE ARGIVE-SICYONIAN SCHOOL.

In spite of the new departure taken by Greek sculpture, the Argive-Sicyonian school under Euphranor and Lysippus remained faithful to its spirit, and continued to translate Nature into vigorous sculptured forms. Lysippus lived at Sicyon, and the period of his most brilliant activity was between 330 B.C. and 320 B.C. Pliny estimates the number of his statues at five hundred ; it is certain that he was extremely productive, which may be explained in part when we remember that most of his statues were in bronze : work in the mould being much more rapid than in marble. Besides this, Lysistratus, brother of Lysippus, had invented the art of moulding in plaster, and had furnished art with a process that might be used

* For example, the Mars and Venus in the Villa Borghese.

in modelling from a cast taken directly from the
human face. This was only a step further in that
path of realistic accuracy in which the school of
Sicyon had ever travelled.

The works of Lysippus included statues of gods,
of heroes, above all, portrait statues of athletes and
of famous persons ; only one allegorical figure by him
has been mentioned, that of *Kairos*, or Opportunity,
which inspired Poseidippus with a pretty epigram,
translated by Ausonius. By his native bent, as well
as by the tradition of his school, Lysippus was
an observer and a realist. He aimed at the expression of the individual type, especially in subjects
where he portrayed the beauty and vigour of the
human body. His Zeus of Tarentum was a colossal
work, which Fabius Verrucosus was unable to trans-
port to Rome because of its huge size.

The legends of Heracles furnished him with numerous subjects ; he represented the labours of the hero
for the city of Alyzia, in Acarnania. Among his
works were four statues of Heracles, which are known
to us both by descriptions and by copies.

The Farnese Hercules of the Naples Museum,
signed by Glycon the Athenian, seems to be the
copy of an original of Lysippus. An inscription
designates as "a work of Lysippus," *i.e.*, a copy, a
statue of Heracles found on the Palatine, and taken
to Florence ; the hero is leaning upon his club in a
pensive attitude. But the inscription, which is certainly modern, deserves no credence. At Tarentum
a statue of Lysippus shows Heracles in repose, seated

O

on his lion's skin, as he is also represented on an antique gem. This statue was, many years later, transported to Constantinople, where it decorated the Hippodrome. An epigram in the Anthology describes the statue as one in which the hero is disarmed by Eros, a subject frequently used by the engravers of precious stones, and appearing on some gems from Florence.

Portraits, in particular those of Alexander, formed an important part of the works of Lysippus. He represented the King of Macedon "in numerous statues, from his childhood up."* Ancient writers mention three principal ones where the king is figured : (1) with spear in hand ; (2) at Granicus, with his friends and his guards, both mounted and on foot; (3) hunting a lion. Among the portraits left us of Alexander, there is not one that may certainly be recognised as a copy of an original of Lysippus ; perhaps the statue in the Louvre, from Gabii, where Alexander stands erect, with a helmet on his head, is a replica of one of the portraits mentioned.

The Museum of the Vatican possesses the copy of the athlete with strigil (*Apoxyomenus*), which was transported to Rome, and placed by M. Agrippa before his baths. We may here better appreciate the style of Lysippus, and observe the alterations he introduced in the canon of Polycleitus.† The artist of Sicyon modified the proportions which had prevailed since the Doryphorus of the Argive master ; he made

---

* Pliny, *N.H.*, xxxiv. 63.

† *Cf.* the notice by M. Collignon in Rayet, *Mon. de l'Art antique.*

the body more slender, the head smaller, and introduced an extreme care in the treatment of the hair.
Tradition was ever powerful in the school of Sicyon ;
Lysippus was the pupil of Polycleitus, and his own
disciples strictly followed in his path. Daïppus,
Boëdas, and Euthycrates, his sons, and Phanis and
Eutychides, his pupils, applied the new canon introduced by their master, and under the impulse given
by the sculptor of Sicyon the taste for colossal compositions continued to prevail in the art of the following period.

At the same time the character of Hellenic art
is about to change. The knowledge of the schools,
however subtle it might be, could not take the place
of the bloom of youthful beauty, which is brilliant in
works of the fifth and fourth centuries B.C. ; the
period of decline is at hand, and in its attempts after
the magnificent and imposing, Greek art more and
more departs from the simplicity and sincere inspiration which made her greatness.

CHAPTER VI.

SIXTH PERIOD.

HELLENISTIC ART.

FROM THE ONE HUNDRED AND TWENTIETH OLYMPIAD TO
THE ROMAN CONQUEST, THE ONE HUNDRED AND
FIFTY-EIGHTH OLYMPIAD (292 B.C —146 B.C.)

ONE of the chief characteristics of this period is the
dispersion and diffusion of the schools. Art changes
its residence, as it were, and Asia Minor becomes the
centre of its activity. Sculptors come into the service
of the Macedonian dynasties, which had divided the
empire of Alexander, and adorn the capitals of Asiatic
sovereigns. Ancient writers have preserved to us
descriptions of famous festivals in these royal courts :
the festival of Adonis in Alexandria under Ptolemy II.,
given in honour of Arsinoë ; that of Antiochus IV.
Epiphanes.* Artists were put under contribution for
these magnificent solemnities ; it was necessary to
work rapidly and according to the taste of the sovereign. Thus Greek art, without ceasing to be fruitful,
is subjected to new conditions, among which are to

* Theocritus, *Idyl.* xv. ; Athenæus, *Deipn.* v. 96 A and 194 C.

be reckoned the individual tastes of its powerful patrons.*

Three schools became the most important homes of art in Asiatic Greece, which had not needed to wait for the conquests of Alexander to become thoroughly impregnated with Greek civilisation : these were the schools of Pergamum in Mysia, of Rhodes, and of Tralles.

§ I. THE SCHOOL OF PERGAMUM.

The school of Pergamum had no ancient traditions ; it was, in truth, only the united company of artists who worked for the Attalidæ, the kings of Pergamum ; the chief of these were Isigonus, Phyromachus, Stratonicus, Antigonus, and Niceratus, author of a votive monument consecrated at Delos in honour of the prince Philetærus. The artists of Pergamum preferred to treat subjects taken from the victories of the Pergamene princes, Attalus II., Eumenes, and Philetærus, over the Galatians, a Celtic people settled in Asia Minor, which disturbed Mysia by frequent invasions. Such was the principal motive of a monument given by Attalus to Athens, which was placed on the south-eastern edge of the Acropolis above the Dionysiac Theatre. The sculptures that decorated it represented the Gigantomachia, the combat between the Athenians and the Amazons, the battle of Marathon, and the defeat of the Galatians in Mysia.

* Decree of Aptera offering to Attalus, King of Pergamum, "a statue on foot or on horseback, according to his choice."—*Bull. de Corr. hellén.*, 1879, p. 425.

According to Brunn, these four groups, including about fifty figures, were ranged in tiers upon a substructure formed of several stories, and were copies of sculptures existing at Pergamum. They were transported to Rome, and fragments of them are scattered through our museums, in Venice, Naples, at the Vatican, and in Paris.* Some of the figures represent fallen Gauls pierced with sword thrusts ; others, Amazons and Asiatics, in their national costume, defending themselves and fighting. The characteristics of the composition show that these groups were a reproduction, on a smaller scale, of larger original works. These smaller reproductions were, however, from the hand of artists of Pergamum.

Two marbles of larger proportions certainly belong to the same order of subjects, and Brunn does not hesitate to regard them as originals. One of them, in the Capitol at Rome, has long been known as the Dying Gladiator, but the twisted collar about the neck of the wounded and sinking warrior, his long hair, his moustache, his type of face, clearly indicate a Gaul. The group in the Villa Ludovisi, wrongly styled Arria and Pætus, represents a Gaul thrusting his sword into his neck after having slain his wife.

Until the last few years these marbles alone represented the school of Pergamum. Excavations made since 1878 by Humann at Bergama, on the site of the ancient Acropolis of Pergamum, have given us a rich

---

* Benndorf has discovered that a statue in the museum of Aix belongs to the group of the Attalidæ.

series of original works, which quite reconstruct this portion of the history of Greek art. These are the sculptures that decorated the great altar dedicated to Zeus and to Athena by King Eumenes II. (197 B.C.—159 B.C.).*

This altar was placed on an immense quadrangular substructure; a flight of steps, cut deeply in one of the faces of the substructure, led to a platform, around which ran an Ionic colonnade backed by a wall. The altar was thus surrounded on three sides by an enclosure open to the sky, forming, as it were, a vast hall. Besides statues placed on the colonnade, two friezes composed the sculptural decoration. One of these friezes extended along the inner wall of the colonnade near the altar ; the other, entirely external, decorated the faces of the substructure.

The former, which was much the smaller, represents the myth of Telephus, the national hero of the people of Pergamum. In the present state of the marbles, it is impossible to reconstruct the composition as a whole,† or even to explain the meaning of all the scenes represented. We may recognise, however, with sufficient certainty, some episodes in the history of Telephus : for instance, the scene where workmen are preparing the skiff designed for Auge, the mother of the hero ; the scene where Telephus is suckled by a fawn in the presence of his father

---

* See *Die Ergebnisse der Ausgrabungen zu Pergamon*, A. Conze, C. Humann, R. Bohn, etc. Berlin, 1880—1882.

† Conze has charge of the restoration of the marbles of Pergamum, which are preserved in the Berlin Museum.

Heracles; finally, that of Telephus with the infant
Orestes before the domestic altar of Agamemnon.

The subject of the Great Frieze, which measures
eight feet in height, is the Gigantomachia. This
vast composition extends around the faces of the
great substructure, covering even the wall on both
sides of the stairway. The marble facings diminish
in height as the stairs ascend; thus Fig. 64 shows
clearly the cuttings for the steps on one of these
facings. The scene is imposing in its effect; Olympian
gods are struggling with giants, some of whom have
serpents' tails, while the younger have the human form;
a furious combat is taking place; draperies are flying,
bodies are intertwined, and the serpents forming
the termination of the giants' thighs twist and unroll
themselves, biting at the shields of the gods with im-
potent fury.

Two fragments in particular are admirable in style:
those which represent Zeus and Athena struggling with
the giants. Zeus, the upper part of his body naked,
advances proudly, having just smitten one of his
adversaries with his thunderbolt; his eagle is fighting
at his side, and one of the giants raising his arm,
covered with the skin of an animal, threatens the
sovereign of the gods. On the other side, Athena has
seized a winged giant by the hair, and drags him off
with a rapid step, while a Victory is flying in the
field of the bas-relief as if to meet her.

Are these sculptures the work of Isigonus, as we
might infer from the fragment of an inscription found
in the excavations? Whoever may have been their

author, they reveal a totally new style in the history
of Greek sculpture ; a violent, impetuous art, attended
by wonderful dexterity in execution. Nothing is
further removed from the meditative art of the fifth

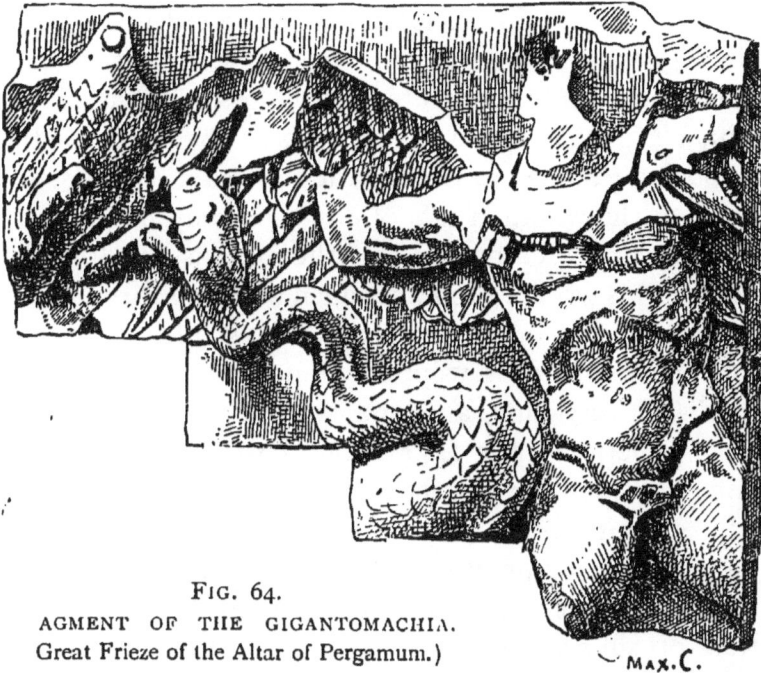

FIG. 64.

AGMENT OF THE GIGANTOMACHIA.
Great Frieze of the Altar of Pergamum.)

century B.C., or from the sensuous grace of the fourth
century B.C. ; it exhibits an almost modern sentiment,
an effort after the new without following any school.
The art of Pergamum is studied and skilful, but at
the same time thoroughly individual ; the wealth of
inspiration in Greek genius is found again in its com-
pleteness, at once varied and powerful.

§ 2. THE SCHOOLS OF RHODES AND OF TRALLES.

The sculptors of the Rhodian school, known to us from ancient writers and from inscriptions, are very numerous. Inscriptions show the great favour in which artists were held among the Rhodians, who received them with honour, and conferred upon them the highest privileges of the city.* One of these artists, Chares of Lindus, executed the colossal statue of the sun, a gigantic figure which betrays the taste for the imposing peculiar to the Asiatic schools. The master-piece of the Rhodian school was the group of Laocoön and his sons,† sculptured by Athenodorus and Agesandrus. This work, which was much admired through the influence of Winckelmann and his school, continued the traditions of Lysippus. The modelling of the torsos is very fine, and the knowledge of the nude form is perfect ; the expression of pain, both physical and emotional, is carried to the extreme. Laocoön contorts himself in most violent agonies, and his sons gaze with horror upon their father. It would be difficult for sculpture to go further in the representation of suffering. The group of the Laocoön is the basis of the famous study of Lessing, in which the celebrated German critic attempts to fix the limits of the arts of painting and sculpture.

With the school of Rhodes is closely connected that of Tralles, known principally in two artists, Apollonius and Tauriscus, authors of the group styled

* Foucart, *Inscriptions de l'île de Rhodes.*
† Restored in modern times by Giovanni Montorioli.

the Farnese Bull.* The group represents the punishment inflicted upon Dirce by the sons of Antiope,
Amphion and Zethus, in punishment for the cruel
treatment received by their mother. The only
ancient unrestored parts of this enormous group are
the torsos of the two brothers, and the lower part of
the body of Dirce, in
which may be recog-
nised the great skill of
the school; the other
parts are only the re-
storations of Giovan-
Battista della Porta,
executed about 1546.
The general effect of
the ancient group, as
inferred from a cameo
from Naples, from coins
of Nacrasa and of Thy-
atira, and from ivories
of Pompeii, was more

FIG. 65.
THE PUNISHMENT OF DIRCE.
(Imperial coin of Thyatira).

simple; the group should, then, be placed in the
series of estimable works where skill in execution
atones for poverty of sentiment.

The period of the conflicts preceding the capture
of Corinth and the final reduction of Greece to a
Roman province is full of obscurity. Pliny declares
that art was in its decline after Olympiad CXXI. (296
B.C.). About Olympiad CLVI. (156 B.C.), however, it

* Brought from Rhodes to Rome by Pollio, and discovered there in
the sixteenth century by Pope Paul III. (Farnese).

had a sort of Renaissance in the Attic school. But it is principally to Rome that we should follow those artists of the new Athenian school, to whom we owe a large number of the statues that adorn our museums. Apollonius, son of Nestor, author of the beautiful Belvedere Torso of the Vatican; Cleomenes of Athens, whose name is affixed to the Venus di Medici; Glycon, Salpion, and Sosibius, represent with talent the artistic traditions of a race that had not yet ceased to be wonderfully endowed; but these names belong to the history of art under Roman domination. The Greeks as a people no longer existed with that public and private life which had exercised so powerful an influence on the development of their genius. We can no longer obtain from the monuments of their art that which is of chief interest to us—the history of their ideas and sentiments.

CHAPTER VII.

STELÆ AND VOTIVE SCULPTURES.

IN this rapid survey of the monuments of Greek sculpture we have considered the style of the schools chiefly in chronological sequence. But this is by no means the only aspect under which Greek marbles present themselves to us. Art, among the Hellenes, found its place everywhere; in private and in religious life, in political affairs as well as in public worship and cultus. The marbles are still to-day the most important monuments whereby we may understand the religious conceptions that nourished popular beliefs, the cultus of the dead, in a word, all that inward life in which lived the soul of the masses. Studied from the point of view of their object and classified by subjects, these marbles form an extended series, the more important of which are the following :—

1. STELÆ ON TOMBS.
2. VOTIVE (*ex voto*) SCULPTURES TO DIVINITIES.
3. HEADINGS OF DECREES, AND OTHER MARBLES RELATING TO POLITICAL LIFE.

§ I. STELÆ ON TOMBS.

Stackelberg, *Gräber der Hellenen*, 1837 ; Pervanoglu, *Die Grabsteine der alten Griechen*, 1865 f. ; A. Conze, *Ueber griech. Grabreliefs*, 1872, and *Bericht über d. vorbereit. Schritte zur Gesammtausgabe der griech. Grabreliefs*, 1874 ff.

Among the Greeks it was exceptiona to have sepulchral chambers. A street with tombs on either

side took the place of the subterranean necropolis.   Thus at Athens, the roads in the outer
Ceramicus were bordered with tombstone monuments which recent excavations have brought to our
knowledge.

These sepulchres were very diverse in character:
sometimes a cavity excavated in the rock and closed
with a slab; at other times a stone sarcophagus, or a
simple grave lined with tiles.

The place of burial was indicated by a monument,
the form of which varied according to the wealth of
the family bereaved.   The most modest were mere
round cippi.   A second type, more ornate, was that
of an oblong stele decorated with rosettes and surmounted by a rich anthemion, where were spread out
the leaves of the acanthus and of the palm; an inscription cut in the marble gave the name of the
dead, that of his father and his tribe (Fig. 66).
Sometimes the stele assumes the form of a vase
decorated by reliefs.   The most carefully-executed
and the richest of the Attic stelæ are constructed
in imitation of a small edifice, and the bas-relief
is enclosed within upright pilasters and the interior
mouldings of a pediment.

We cannot specify all the types of stelæ followed in Greek countries; they were of infinite
variety according to local usages.   The tombs of Asia
Minor carved in rock, with actual façades like temples,
had nothing in common with the elegant stelæ of
Attica, or with those of Bœotia and of Thrace.   We
will restrict ourselves to Attic stelæ, and will indicate

ΕΓΙΚΡΑΤΗΣ
ΚΗΦΙΣΙΟΥ
ΙΩΝΙΔΗΣ

FIG. 66.—STÉLE AT ATHENS.

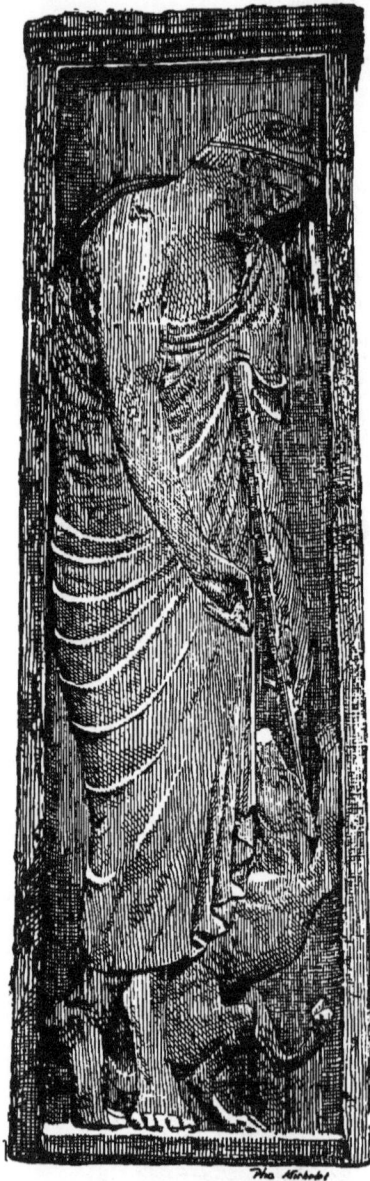

FIG. 67.
STELE FROM ORCHOMENUS.

the principal motives followed in their sculptural
decoration.

The Dead and his Family. — According to
most ancient usage, the
sculptured and painted
stele presented a likeness
of the departed. Thus
the stele of Velanideza,
signed Aristocles, shows
a Greek warrior in armour
(see Fig. 41); on another,
found at Orchomenus,
signed by Alxenor, the
dead man, wearing a
chlamys, is playing with
his dog. Accessories often
suggested favourite occu-
pations; on an archaic stele
of Lyseas, which is painted
but not carved, the painting upon the pedestal
represents a rider mount-
ing a race-horse. For a
long time these figures
were executed in advance
of their use, and offered
no individual features,
and were not portraits.
The inscription alone gave

the monument its definite application, and the attempt to reproduce upon a monument the features of the dead does not assert itself before the Græco-Roman period. The simple representation of the dead is often made complex by accessory figures; here domestic animals, there a bird; while in order to suggest the games of the palæstra to which the dead was addicted, a child standing near holds a flagon of oil and a strigil. More complete still is the representation of a person surrounded by the members of his family, as in the charming stele of Protonoë, where the young dead girl is figured in the midst of her relatives.

Scenes from common life.—In order to diminish the feeling of sadness at the thought of death, the Greeks loved to represent the dead in the midst of objects which had busied or pleased him in life. Thus several stelæ show an Athenian lady at her toilette, with her women, who hold before her a jewel-casket; this is the subject of the stele of Hegeso (Fig. 68), one of the best marbles from the Ceramicus. The same subject is treated on Athenian vases with white background, but with special details which make us understand the spirit of these representations. Here the dead woman is figured as if seated in front of the stele; it is as if her image had been detached, in order to restore her to the living, and to show her engaging in familiar occupations. To the same spirit are due the figures of persons on horseback, running, or hunting. At times the motive is in allusion to some definite fact, as in the monu-

P

ment of Dexileos ; the representation of the fighting knight calls to mind his glorious death before the walls of Corinth.*

The Farewell.—In this scene the dead receives the farewells of his nearest relatives, who extend their hands towards him with an expression of sadness. The word χαῖρε ("farewell") is often added to the inscription, and expresses the sorrow of survivors. The meaning of this scene has often been discussed. Ravaisson believes that there is to be seen in it, instead of separation, the reunion of the dead with his family in the other world.†

A person on a rock with a boat near at hand.—The best known instance of this motive is the stele of Glycon, in the museum at Athens ; it doubtless suggests a shipwreck, or death at sea.

Burial.—This scene is of rare occurrence, but the specimens of it have a great value, in showing that the Greeks did not avoid the representation of death, and that the subjects mentioned above do not necessarily suggest scenes of the future life.

The Funeral Banquet.—This subject, which is very frequent in Attica, is of great interest in the study of funeral customs. It has several variations, but the most complete scenes represent the dead reclining on a banqueting couch ; before him stands a three-legged table ; he receives from his relatives the food necessary

---

* Inscription : "Dexileos, son of Lysanias, of the deme of Thoricus, . . . one of the five knights slain at Corinth." [*Cf.* Hicks, *Greek Hist. Insc.*, p. 125.]

† *Le Monument de Myrrhine* in *Gazette arch.*, 1879.

FIG. 68.—STELE OF HEGESO.
(Ceramicus, Athens.)

to maintain the half-material life that he lives in the tomb. The scene is illustrated by the custom of νεκύσια, actual funeral banquets which, celebrated at the tomb, were popularly believed to be of benefit to the dead himself.

The scenes that we have just indicated originated in modes of thought peculiar to the Greeks, and the commentary upon these thoughts is furnished rather by monuments than by ancient writers. There are other scenes arising from better known mythological beliefs, the interpretation of which offers fewer difficulties. Thus on one marble from the Ceramicus we may see Charon and his boat; on a beautiful vase decorated with reliefs, Hermes, conductor of souls, is leading a young girl, named Myrrhine.*

Another series shows the dead man made a hero, standing by his horse, and in the field of the stele a serpent, symbol of divine presence, coiling near a tree. Again the stele has upon its face representations of mythological beings, like the Sirens, muses of death, and the Harpies, who carry away souls. Finally, in the Græco-Roman period, when Oriental superstitions had gained general credence, tombstone subjects were drawn from the Dionysiac cycle. Bacchic emblems seem to promise to the dead a renewal of his

---

* Ravaisson, in the article cited above, sees here a reunion in Elysium represented. O. Benndorf, *Mittheil. des deutsch. arch. Inst.*, 1878, believes that living persons are to be seen in it, filled with a religious awe in the presence of the young Myrrhine, carried away by Hermes Psychopompus.

existence in a happy life, "in the midst of the com-

FIG. 69.—FUNERAL BANQUET.
(Commonly called "The Death of Socrates.")

pany of Bacchus, of the Satyrs, and of the Naiads."*

* Inscription of Doxato ; Heuzey, *Mission de Macédoine*, 1864—1878.

While giving full weight to the mythological allusions, and to the changes which the types illustrated have suffered in process of time, the ruling idea of these Greek stelæ seems to be to represent the dead with that which he loved in life. Apart from the interest aroused by the monuments in the study of beliefs, they often commend themselves for their exquisite elegance of execution. The marbles of the fourth century B.C. show how profoundly the great schools of art had made themselves felt even in works of minor importance. On the most beautiful Athenian stelæ the excellence of the plastic form and the restrained and calm expression of sentiment are well adapted to make us understand what are the finest qualities of Atticism.

§ 2. VOTIVE MARBLES DEDICATED TO DIVINITIES.*

The subjects sculptured on votive marbles are infinite in variety. Hellenic religion gave a large place to ceremonies, to visible demonstrations ; and in the religious life of the Greek there was many an occasion when his devotion to a god expressed itself in a votive offering—a marble slab ornamented with reliefs, which was dedicated in the temple enclosure. We cannot here enumerate in detail all the representations figured on these marbles : we will restrict ourselves to the more important—to those that have most attracted the attention of archæologists.

* Stephani, *Der Ausruhender Heracles*, 1854 ; Welcker, *Alte Denkmäler*, Vol. I., and Schöne, *Griechische Reliefs*, 1872.

FIG. 70.—VOTIVE MARBLE TO ASCLEPIUS AND HYGIEIA.

Votive offerings to Asclepius and to Hygieia, the divinities of health, form a numerous series, lately much enlarged through recent excavations at the south-east of the Acropolis, on the site of the temple of Asclepius.* These bas-reliefs, placed in the temenos of the temple here and there in the sacred enclosure, represented varied scenes ; they may be grouped into the following classes :—(1) of simple adoration ; (2) of offerings and sacrifices ; and (3) of banquets. In the simplest scene, the family has just offered prayers to Asclepius and Hygieia, whose figures are of very large size. Elsewhere the family has just sacrificed a victim on an altar, a sow or a ram, in the presence of the two divinities. The banquet scene is more complex ; Asclepius and Hygieia are seated upon a couch, before a table covered with offerings, while the suppliants are facing them.† This scene has sometimes been identified with that of the funeral banquet, but the fact that similar marbles have been found on the site of the temple of Asclepius makes this opinion untenable.

Another series of marbles comprises votive offerings to Serapis and Isis. This class has not always been kept distinct from the foregoing ; but the *polus* or headdress worn by Serapis shows that some other divinity than Asclepius is indicated. These marbles belong to a late epoch, when purely Greek ideas had

* *Cf.* P. Girard, *L'Asclépieion d'Athènes d'après de récentes Découvertes*, 1880.

† The presence of the horse, the head of which appears at a window, has not yet been satisfactorily explained.

received an admixture of Egyptian beliefs, and when the Osiris of Egypt had become assimilated, under the name of Serapis, to the Pluto of Hellenic religion.

Votive marbles to Heracles, to Pan, and to the Nymphs, etc., are less numerous, and an analysis of their characteristics would carry us beyond the limits set to this sketch.

§ 3. MARBLES RELATING TO POLITICAL LIFE.*

The political acts of the Greeks were recorded on stelæ, and often, above the inscription, was placed a bas-relief which indicated in allegorical form the act mentioned in the decree. These monuments deserve careful study, for they show how the Greeks personified abstract entities, such as the Republic of Athens, the Senate, the People, and even political and administrative qualities. According to the nature of the decrees, the accompanying headings of the stelæ may be classified as follows :—

1. *Treaties of Alliance.*—Cities forming parties to a treaty are often represented by the features of their tutelary divinities. On a decree relating to an alliance between Athens and Neapolis, the Athenian State is represented by Athena, and Neapolis by Artemis Parthenos, worshipped in that city ; the two goddesses are shaking hands. In other cases, the people of Athens itself, the *Demos* introduced upon the dramatic stage by Aristophanes, is figured as a

* Schöne, *Griechische Reliefs ;* A. Dumont, *Bull. de Corr. hellénique* II., 1878, pp. 559—569.

man wearing a mantle ; thus above a treaty of alliance with Corcyra, we see the demos offering his

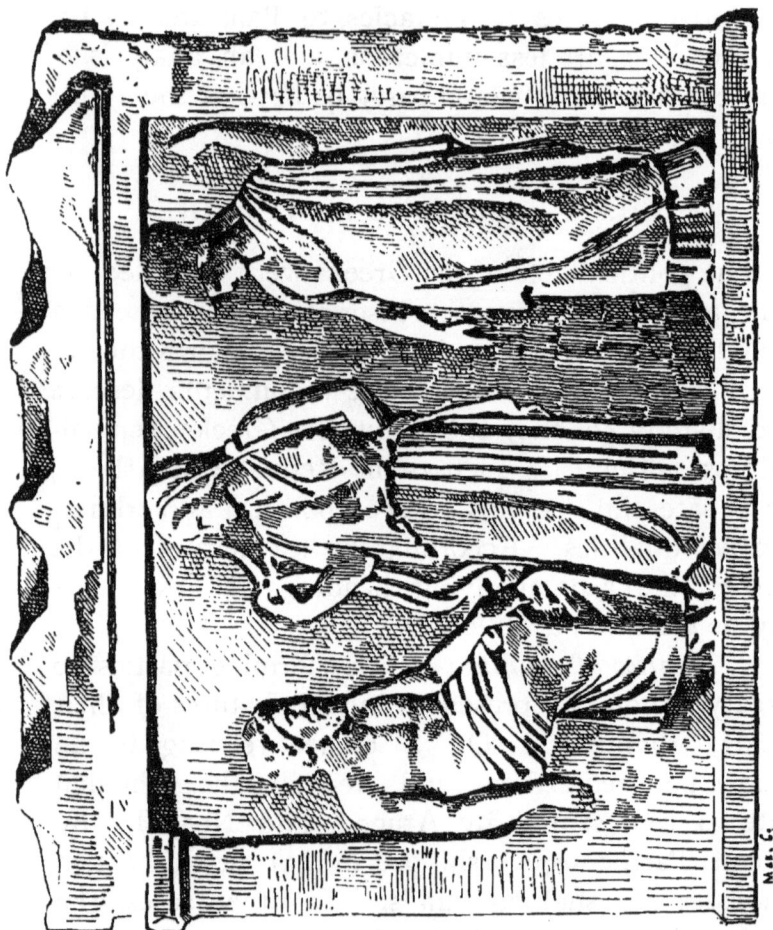

FIG. 71.—HEADING OF A TREATY BETWEEN ATHENS AND CORCYRA.

hand to the friendly city, personified in a young woman, while Athena presides over the scene. (Fig. 71.)

2. *Laudatory inscriptions in honour of States or of individuals.*—When the Republic wished to reward a citizen of a foreign city, it conferred upon him the title of *proxenus*, together with the honour of a crown.

FIG. 72.—HEADING OF A DECREE CONFERRING PROXENIA.

This subject is shown in a certain number of bas-reliefs, where Athena, wearing her helmet and ægis, is crowning the honoured person. These allegorical motives were also treated in sculpture on a larger scale. Ancient writers mention a monumental group, representing the Demos of Athens, crowned by Byzantium and Perinthus.

3. *Records of finance.*—These marbles show principally the personifications of Athena, and of the branches of the body politic. It is not unusual to find the Senate (Βουλή) represented in the likeness of a woman, to whom a goddess offers her hand as a sign of satisfaction.

4. *Accounts of liturgies, magistracies, etc.*—One of the most interesting of these reliefs represents not the State nor the body politic, but a moral quality. Here the goddess of good management, *Eutaxia*, crowns a person in congratulation upon his having creditably acquitted himself in a liturgy.

We have mentioned only the most important examples. Studied in their details, these marbles reveal with what simple art at the best period abstract entities were conceived. There is no far-fetched allegorical suggestion, no multiplication of confusing accessories ; the artist confined himself to clothing pure abstraction with a plastic form, and attempted not so much to explain an allegory as to create a work of true art.

The museums of Athens possess many more sculptured monuments that are closely connected with the political life of the Athenians. Of this number are the marbles relating to the *Ephebi*, the bas-reliefs upon which show the games and exercises of young Athenians while in the direct care of the State ; on some of them the ephebi are crowning their magistrates and other minor officers, as, for

FIG. 73.—BUST OF THE COSMETES OF THE ATTIC EPHEBI.
(From the *Bull. de Corr. hellénique.*)

instance, the *cosmetes*. A special class of these marbles is formed by the busts of the officers of the ephebi, which were placed upon tetragonal stelæ, after the manner of Hermæ, upon which was carved the decree. These are actual portraits, and acquaint us most accurately with the Greek type of the already advanced epoch in which they were executed.

We can easily see, from the hints already dropped with regard to these classes of monuments, the place held by plastic art in the private, religious, and public life of the Greeks. Beliefs, ideas, political acts, assume a visible and tangible shape by means of them. The study of these marbles may, perhaps, reveal to us facts that would absolutely escape our knowledge were we reduced to the testimony of the ancient writers alone.

* A. Dumont, *Bull. de Corr. hellénique*, I., pp. 229—235, and II. Plates VI., VII.

Book Fourth.

Terra-cotta Figurines.

THE systematic study of these diminutive monuments, which form an important chapter in the history of Greek ceramics, is of very recent origin. Only within a few years has it taken a distinct position in archæology in general, and only lately have the terra-cotta collections in European museums been described and classified. The art of the *coroplastes*, or modeller of figurines, is one that should not be overlooked. It reveals to us, often in an unexpected aspect, the familiar and popular side of ancient life, the secret of which is not to be found in the higher arts. The study of terra-cotta products forms a natural transition from the history of sculpture to the study of painted vases, the latter constituting the more important branch of ceramics.

It is unnecessary to remark that the art of the modeller in terra-cottas was employed in a great variety of forms. From his workshop came many objects destined to adorn monuments: gutters in painted earthenware, gargoyles adorned with masks and heads, as the "main tiles in the form of lions' heads" mentioned in the inventories of the Arsenal of the Peiræus, specimens of which have been found.

We shall now consider the work of the *coroplastæ**—
"doll-modellers," as they were popularly designated
—whose works furnish us with an indispensable commentary on the monuments of sculpture on a grander
scale.

Terra-cottas form two distinct categories: stamped
plaques and figurines.

§ I. STAMPED PLAQUES.†

About fifty of these monuments in different
European collections are known at the present date.
The process of manufacture was uniform. They were
executed by means of a matrix containing the design,
which was placed upon a thin cake of moist clay. After
the impression was thus obtained, the workman carved
the details of the plaque, finishing the contours of
the figures and supplying the background, in order
to give more lightness to the work. After firing, the
plaque was painted, and then was ready to be applied
to the interior wall of a house or of a tomb.

The place of manufacture of these productions is
not certain. Because many of them have been found
in Ægina, at the Peiræus, at Melos, it by no means
follows that workshops for their production were
established in these places. There is reason to
believe that their manufacture was restricted to a

---

* Κοροπλάσται. "Coroplastes : one that models figures of living
beings, not only images of little boys or of little girls, but any kind of
figure."—*Etym. Magn.*, p. 530.

† Schöne, *Griechische Reliefs*, plates xxx.—xxxv. ; O. Rayet,
Monuments de l'Art antique, 1er livraison, pl. x.

Q

Fig 74.—Plaque in stamped terra-cotta.
(Representing a funeral procession.)

few localities, and the style of these plaques, which have all the characteristics of the art of the fifth century B.C., leads us to believe that this manufacture was of short duration.

The subjects illustrated on these plaques are sometimes mythological, sometimes scenes from ordinary life. Two of the most beautiful specimens of the former variety are a stamped plaque of Melos, representing Bellerophon fighting the Chimæra, and Perseus slaying the Gorgon, whose soul escapes from the decapitated body in the form of a little figure. The struggle between Thetis and Peleus, on a plaque found in Ægina, is of great interest from the point of view of composition; we discover in it the slender and somewhat lank proportions of late archaic art. The museum of the Louvre possesses a very beautiful specimen, showing Orestes and Electra near Aga-memnon's tomb, subjects frequently reproduced on monuments of this nature.

Subjects taken from daily life are less numerous, but these plaques are of infinite value in the study of ancient life, manners, and customs. Such is a stamped plaque representing a funeral procession (Fig. 74): the dead body, placed in a small waggon drawn by two horses, is accompanied by friends and mourners on the way to burial (ἐκφορά). The persons forming this procession are those authorised by law; the *enchytristria*, carrying upon her head the vase for libations; the flute-player, and members of the family, the women with unbound hair; young men in military dress, but unarmed, seem to reproach the dead for having left them. This plaque is, as it were, an ocular commentary on passages in ancient writers and in ancient laws that have come down to us relating to this subject.*

Interesting as are these plaques in the subjects they illustrate, they are equally interesting on account of their execution. In general there is a certain uniformity of style; great simplicity in modelling, no very high relief; somewhat of stiffness in the attitudes. It is unquestionably the youthful and naïve style of the first years of the fifth century B.C. that we here meet, the tradition of which may have prevailed longer among the modellers in clay than in the schools of sculpture.

§ 2. TERRA-COTTA FIGURINES.

HEUZEY: *Nouvelles Recherches sur les Terres-cuites grecques,* and *Les Figurines antiques de Terre-cuite du Musée du Louvre,* 1878 ff.

KÉKULÉ: *Griechische Thonfiguren aus Tanagra,* 1878.

O. RAYET: *Les Figurines de Tanagra au Musée du Louvre* (in *Gazette des Beaux-Arts*), 1875.

J. MARTHA: *Catalogue des Figurines en Terre-cuite du Musée de la Société archéologique d'Athènes,* 1880. (At p. xxix. will be found a complete bibliography of the subject.)

In the study of figurines, the question as to the place of production is one of the most important; it furnishes the safest datum as to classification. From this point of view, the portions of the Greek world that supplied most monuments of this kind are, in Greece proper, Attica, Bœotia, and especially the town

* Decree of Iulis in Ceos, *Mittheil. des deutsch. arch. Instituts in Athen,* I., p. 239.

Q 2

of Tanagra, Locris, Peloponnesus, and the Cyclades; in Africa, Cyrenaica; in Asiatic Greece, Rhodes, Ephesus, Pergamum, and Tarsus. Of the localities where figurines have been found the best known are those either where regular excavations could be conducted, as in the necropolis of Cameirus in Rhodes, or those where the products can be systematically studied, as is the case with the terra-cotta figurines from Tanagra. The essential point in the study of terra-cottas will long remain this matter of their provenance, or place of production; only by examination of the clay, and of the processes of manufacture, shall we be enabled to group them according to localities, and to apply to them the method employed in the study of modern ceramics.

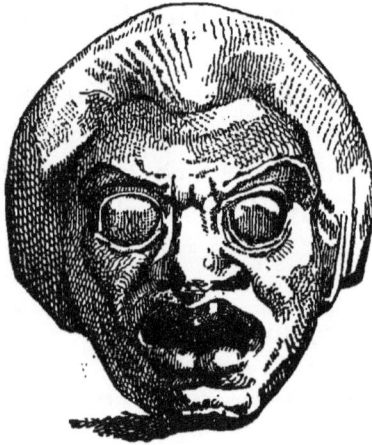

FIG. 75.—TRAGIC MASK.
(Terra-cotta.)

From a chronological point of view, classification of terra-cotta figurines presents certain difficulties. Among those of most ancient style, some certainly are contemporaneous with the beginnings of Greek art; but, as these types never ceased to be reproduced with scrupulous fidelity in order to satisfy popular demands, the archaic style is no infallible indication

of antiquity.   The date even of the Tanagra figurines
cannot be fixed with any definiteness ; in general they
belong, both in style and in composition, to the fourth
century B.C., or to the first years of the third century
B.C.   The figurines from Tarsus, finally, are not
anterior to the Seleucidæ, and mark the latest epoch
in figurine manufacture.   While making these reserva-
tions, and while admitting at once that terra-cottas do
not strictly belong to the period apparently indicated
by their style, we are able to follow the same variations in style that we have noted in the marbles :
primitive and archaic style, the severe art of the fifth
century B.C., the art of the fourth and third centuries
B.C., and finally the style of the last period, corresponding with the extension of Hellenism after the
time of Alexander.

(1) *Primitive and Archaic style.*—These figurines
are most frequently ordinary idols, and have been
found in great numbers in the necropolis of Tanagra
and in that of Tegea. They recall the old images that
were objects of worship, carved wooden slabs (σανίς)
like the Samian Hera, or wooden statues (ξόανα) like
the Athena Polias in the Acropolis of Athens, coarse
images hewn with a hatchet, that were not forgotten
even in the midst of the most beautiful products of
plastic art.   The technique of these small idols is the
simplest.   They were rapidly modelled by hand in
cakes of clay, cut into rectangular shapes ; two stump-
like appendages were the arms ; the face was suggested
by the workman by pinching the clay in his fingers.
Sometimes the head was modelled with greater care,

and wore the high cylindrical head-dress called the *polus;* accessories like ribbons, diadems, pendants, and necklaces, with which images of worship were covered, were imitated by beads of paste, applied to the moist clay : the toilette of the figurine was then completed by a streaking with red or with bistre. It is rather venturesome to ascribe mythological names to all these rude statuettes; we can safely do it only in rare cases. Gerhard believed that an imitation of the ξόανον of Athena Polias might be recognised in an Attic terra-cotta, because of a sort of an ægis covering the breast of the figurine.*

These terra-cottas have hardly more than an archæological value. Art is more distinctly apparent in the figurines of the archaic epoch, which reflect the style of sculpture of the sixth century. In this number we frequently recognise figures of seated goddesses, with the calm and solemn attitude expressed in the epithet εὔθρονος. Wearing the *stephane* upon their heads, their faces surrounded by the folds of a veil, they hold their arms close to their bodies and rest their hands on the knees, like the statues adorning the sacred way at Branchidæ ; the tomb cities of Rhodes (Cameirus in particular) and of Tanagra have furnished numerous examples. This class of terra-cottas marks the transition from the coarse primitive idols to the figurines of the later period.

(2) *Severe style.*—By this term may be characterised figurines that are the product of the severe art

---

* *Gesammelte akad. Abhandlungen,* I., p. 232.

of the fifth century B.C., where traces of archaism have not completely disappeared. They had a different character from the figurines of the preceding class: they here commonly represent mythological subjects, and figures of divinities, treated with the religious feeling that prevailed at the time of the Persian wars.* Certain types of feminine divinities in particular are marked by an air of austere dignity. Represented as standing with arms hanging down, clothed in costumes with straight vertical folds, they preserved a sculptural and hieratic attitude. Such is a beautiful statue of Core, found at Thisbe, the

* It is not unusual to find traces of fire on these figurines : they were often burned upon the funeral pyre.

FIG. 76.—HERMES CRIOPHORUS.
(Terra-cotta from Thespiæ.)

style of which has all the characteristic features
of the art of the fifth century B.C.   We are right
in supposing that some of these figurines reproduced, in much smaller dimensions, certain celebrated works of the sculptor's art placed in
temples.   A figurine from Thespiæ, for example,
representing Hermes carrying a ram upon his
shoulders, seems to have been directly inspired by the
Hermes Criophorus of Calamis, executed for the
people of Tanagra in memory of a plague that had
desolated the city. "The god turned a pestilence away
from the city by carrying a ram around the walls." *
Other figurines represent Hermes wearing a small
cap (κυνῆ), and carrying the ram under his arm (Fig.
76) : it was thus that the Æginetan sculptors, Onatas
and Calliteles, conceived the statue of the god that
they executed for Olympia.†   The taste and tendency
of the coroplastæ for reproducing works of sculpture
explains the style of certain figurines, which, enlarged
by the aid of imagination, might be regarded as
faithfully enough representing the lost works of some
of the great masters.

To the figurines of the severe style belong busts
moulded in terra-cotta, where may be recognised
types of the Chthonian divinities.   These busts were
commonly pierced in the upper part with a hole to
receive attachments ; they could thus be hung on the
inner wall of a tomb ; they seemed to be rising from
the earth.   In the view of the Greek these images

* Pausanias, ix. 22. 1.
† Pausanias, v. 27. 6.

FIG. 77.—DEMETER.
(Bust of Stamped Terra-cotta.)

called to mind the divinities of the lo\
appearing to mankind by showing only tl
part of their bodies above the earth. One
most remarkable examples of this kind of monι
is a bust of Demeter, published by Heuzey,* prι
from Tanagra. The face of the goddess is, iι ᵣₐιι,
enclosed between the folds of a veil; this rests upon
a scarf or a sort of turban, ornamented with a painted
fretting, that binds the hair, while the two hands are
placed against the breast. By an analysis of the face
and attributes, Heuzey is led to designate this bust as
a Demeter: " I know few antique figures even in
marble where the depth of pensive and reflective
feeling produces upon the spectator an equally pene-
trating emotion."†   Sometimes, always with the same
signification, the statuette is cut half-way down the
body, as may be seen in a terra-cotta from Athens,
representing Core, daughter of Demeter, who was also
a divinity of the lower world. (Fig. 78.)

(3) *Style of the fourth century* B.C.—This was a
new period in the industry in figurines, characterised
principally by the Bœotian manufactures of Tanagra,‡
Thisbe, and Aulis, and by those at Athens and at
Corinth. The process of manufacture is the same as
that employed by the coroplastæ for all figurines
made with a mould. Though statuettes of a single

---

* *Monuments grecs publiés par l' Association des Études grecques*, 1873,
pl. 1.
† *Ibia.*, p. 19.
‡ The museums richest in Tanagra figurines are those of the Louvre
of Berlin, and of Athens.

piece have been found, generally, in the case of the
most finished figurines, the front is first obtained by a

FIG. 78.—CORE.
(Terra-cotta from Athens.)

mould in repoussé ; the head and other extremities
were made separately. This front portion was then
attached to a back, which commonly has a smooth

surface ; the head, hands, and feet are afterwards
added. In the centre of the back is bored a vent-hole
to allow the air and moisture to escape, and the statuette is placed upon a small pedestal. The details of the
head and costume, engraved by means of a pointed instrument, give each figure its individual character and
personal physiognomy ; thus, in the cases of a museum,
it is not uncommon to see two figurines from the
same mould that differ considerably in details.
After the first firing, the statuette passed into the
hands of a workman whose duty it was to paint it
from head to foot ; he sometimes gilded the ornaments, such as earrings, necklaces, and diadems. The
colours most frequently employed are blue, all shades
of pink, red, reddish-brown, and black.

The interpretation of the subjects represented,
which for archaic figurines is an easy matter, is here a
much more delicate task. We cannot enter into
details of the discussions which have principally
been centred on this point. Are we to attribute a
mythological sense to terra-cottas of the fourth century B.C., or are we to see in them, as a rule, only
representations of ordinary every-day life ? Both
propositions have been sustained with a rich array of
arguments ; but, at the same time, the latter opinion
seems to accord best with the character of the types
figured, where a mythological sense is far from being
clearly intended. The sight of these figurines, with
their life-like air and *spirituelle* execution, carries us
rather into the world of living mortals than into the
Hellenic Olympus, and we are constrained to ask of

them the secret of the daily life of the Greeks in all
its most entertaining details.

FIG. 79.—TERRA-COTTA FROM TANAGRA.

On account of their infinite variety, it is impossible
to group these subjects into classes, and we can give

only a rapid glance at the most frequent representations. Subjects taken from the life of men are of
rarer occurrence. In the Tanagra statuettes, sometimes a small boy is figured seated on a sort of
cippus (Fig. 79) ; he holds a bag that seems to con-
tain a ball ; his head is decked with flowers. Sometimes an ephebus is represented, wearing a chlamys,
carrying in his hand the instruments of the palæstra,
emblems of physical exercise, which held so large a
place in the life of the Greeks.

By far the most numerous are scenes in the life of .
women. Here the imagination of the coroplastæ is
inexhaustible, and their inventive genius is accompanied by extraordinary manual dexterity. With
perfect art are they able to vary, by differences of
attitude and of arrangement, the very simple motive
treated by preference, viz., a Greek lady, either in
home dress or ready to walk abroad. The articles in
the costume are few in number ; a long tunic, falling
to the feet, and bound at the waist by a girdle, and a
mantle (*himation* or *calyptra*) of finer material, that
has an infinite variety of details, in accordance with
the taste and caprice of the owner. The Bœotian
figurines show to what uses the modellers would put
these different elements. Sometimes women are completely draped in the himation, and have but a small
part of the face in sight; again, they wear also a little
hat, and in the hand hold a fan in the shape of a
lotus-leaf. At other times, it is a young girl that ha
allowed the calyptra to slip from her shoulders, or
that has so enveloped herself in it that not even her

hands are visible. Some of these figurines have a perfectly modern appearance and general effect; there is nothing even in the details of the way of wearing the hair that would lead one to suspect in them anything but contemporary works of art. Figurines are sometimes grouped in pairs: thus, in a beautiful group found at Corinth, of which different establishments in Greece proper furnished replicas, is exhibited a young girl carrying another girl upon her shoulders—the game of ἐφεδρισμός or of ἱππάς, still played in modern Greece.

Though representations of mythological subjects, like divinities, loves, etc., are not completely neglected by the coroplastæ of the fourth century B.C., subjects from ordinary private

FIG. 80.
GREEK LADY WEARING A PETASUS.
(Terra-cotta from Tanagra.)

life are by far the most frequent. How shall we explain this change in figurine industry taking place within two centuries? It seems that the modifications to which the Hellenic mind and popular religious belief had become subject after the Peloponnesian war had something to do with it. Found in tombs, these figurines, which have a votive character, are evidently connected with beliefs as to death and burial. That in the centuries of belief, like the time of the Persian wars, idols representing gods should be interred with the dead, seems perfectly natural; the dead man must be surrounded by his gods, his arms, his jewels, everything connected with him in life. Later, when the religious sentiment had loosened its hold, the people continued to respect tradition, the meaning in which was obscure; they continued to place with the dead in the tomb figurines that should recall to him in the other life his mortal companions in this. These figures should be the delight of the half-real life that he was supposed to live in the tomb. They took the place of living beings, slaves, horses, which in heroic times had been immolated on the tomb of the dead warrior, in order that he might be accompanied to Hades by his habitual companions.

We have dwelt upon the Tanagra figurines because of their importance and artistic value. They are interesting not only in that they afford information as to Greek life in the fourth century B.C.; they are the product of one of the periods in the development of Hellenic genius in which predominated a more subtle and more refined, though less strictly religious art.

(4) *Style of the third century B.C.*—This period is characterised especially by the manufactories at Pergamum, Smyrna, Ephesus, Myrina,* Miletus, and Tarsus. They present appreciable differences in technique ; the clay is finer and more compressed, and consequently the execution has a hardness not possible in the Tanagra figurines, that were freely retouched according to the fancy of the artist. The subjects, moreover, seem to be often inspired by work of contemporary sculpture, when they are not actually casts made direct from small bronzes. Amid the types of Heracles, of Eros, and of Aphrodite, are found the

* The Louvre possesses a rich collection of figurines from Myrina, obtained from excavations recently conducted by Pottier and S. Reinach.

R

FIG. 81.—BACCHANTE.
(Terra-cotta from Tanagra.)

favourite proportions of the school of Lysippus, the small head and longer body, in order to obtain an effect of elegance and slenderness.

The types most frequently treated by Asiatic coroplastæ are those that the art of the fourth century B.C. had brought into favour, the charming divinities of the cyclus of Aphrodite, of Eros with outstretched wings, imitated, without doubt, from the famous statue of Praxiteles, of Aphrodite Anadyomene, and of Eros and Psyche embracing each other, as in the beautiful gilded group found in Smyrna. These monuments give rise to interesting comparisons with works of sculpture, which, in their careful execution, they closely approach. At the same time imagination and free imitation of nature had their place. From Asia Minor comes a series of interesting statuettes representing grotesque creatures, mountebanks, hawkers, etc., treated with a firm and swift hand, which show us that the Greek spirit understood caricature. (Fig. 82.)

The most recent of Asiatic manufactories appears to be that of Tarsus, the products of which are not anterior to the times of the Seleucidæ. It is known to us principally by fragments, in part in the Louvre and in part in the British Museum, which came from a hillock called Gueuslu-Kalah (Belvedere Fort), near the walls of ancient Tarsus. The excavations were made under the direction of W. Barker, in 1845, and of Langlois, in 1852.* The result of the observations

* See Heuzey, *Les Fragments de Tarse au Musée du Louvre,* in *Gazette des Beaux-Arts,* 1876.

of Heuzey is that these fragments were pieces of
refuse material that had
been injured in the
firing; rejected by the
modellers, and thrown
together, they made in
time a large mass. In
spite of their humble
origin, these terra-cottas
have a real value. They
exhibit all the features
of the prevalent style
after the time of Alex-
ander—that which the
diffusion of Hellenistic
civilisation extended to
all regions of the Greek
world. The modellers
of Tarsus seem to be con-
nected with the school
of sculpture that had
its centres at Rhodes,
Tralles, and Pergamum ;
they copied the works
of these sculptors, and
thus among the frag-
ments in the Louvre is
to be found a group
representing the famous
Laccoön group. At the

FIG. 82.—PEDLAR.

(Grotesque Figurine from Asia Minor.)

same time their style had something of the theatrical

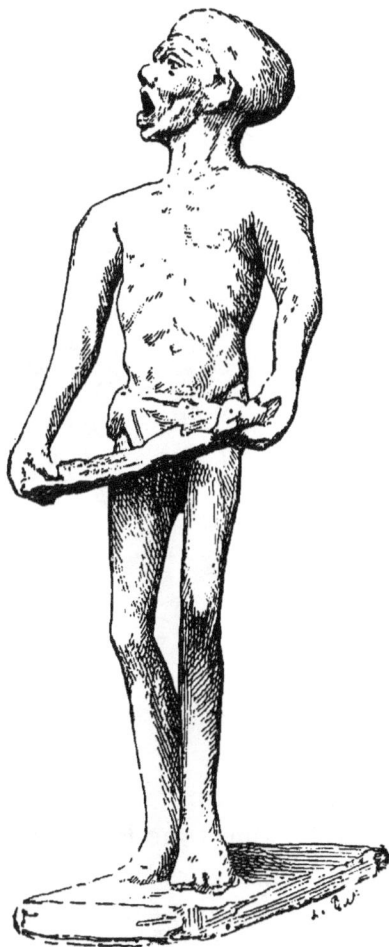

R 2

and affected taste that prevailed under the successors of Alexander.

The subjects figured are equally worthy of attention. They exhibit a curious combination and mixture of types taken from Hellenic mythology, with Oriental attributes. The distinctive character-

FIG. 83.—TERRA-COTTA FRAGMENT FROM TARSUS.

istics of divinities are often mixed and confounded; at the same time local divinities, like Mên and Atys, are often figured. These are most valuable materials for the study of that process of assimilation brought about in Asia Minor after the Greek conquest, which mingled native and local forms of worship with mythological beliefs originating among the Greeks proper.

Book Fifth.

PAINTED VASES.

CHAPTER I.

GENERAL QUESTIONS IN THE HISTORY OF CERAMIC ART.

GERHARD : *Rapporto Volcente*, 1831.
OTTO JAHN : Introduction to his *Beschreibung der Vasensammlung König Ludwig's in der Pinakothek zu München*, 1854.
CH. LENORMANT and DE WITTE : Introduction to their *Élite des Monuments céramographiques*, 1844 *ff*.
DE WITTE : *Études sur les Vases peints*, 1865.
BIRCH : *History of Ancient Pottery*, 2nd ed., 1873.
A. DUMONT : *Peintures céramiques de la Grèce propre*, 1874.
A. DUMONT and CHAPLAIN : *Les Céramiques de la Grèce propre,* 1881 *ff*.
GERHARD : *Auserlesene griechische Vasenbilder*, 1840—1858.

VERY little is known about painting among the Greeks. Descriptions given by ancient writers of the works of Polygnotus, of Zeuxis, of Apelles, of Parrhasius, of Protogenes, etc., are not sufficient to give us an exact idea ; and the paintings of Pompeii, where one finds occasionally a reminiscence of celebrated compositions, belong especially to the history of art in Italy. They betray, besides, the predominant taste of a somewhat degenerate age, which may be termed the Alexandrine or Hellenistic age. In the absence of other monuments, ceramic paintings have a peculiar value. They represent a branch of an art, the principal works of which are lost to us. But not for this only do we take interest in painted vases ; the subjects that decorate them, by reason of their variety, are like an illustrated commentary on all Greek antiquity.   Mythological scenes, where figure all the Olympian divinities, heroic legends, religious and funeral ceremonies, gymnastic exercises, scenes of feasting, of betrothal, of marriage, of the toilette, etc. ; all these subjects are depicted upon the vases, and revive for us all antiquity in its religious and familiar life.   We may judge of the importance of these data by the number of the painted vases that have been preserved.   There are not less than 20,000* scattered through museums and private collections.†

The first painted vases to attract the attention of scholars, about the close of the seventeenth century, were found in Tuscany ; the general opinion was that they were products of Etruria ; hence the designation *Etruscan vases*, which is now universally discarded.   It is not necessary to examine all the erroneous theories to which the interpretation of Greek vases gave rise during the eighteenth century.

---

* Birch, *History of Ancient Pottery*, p. 149.

† Museum of the Louvre, and Cabinet des Médailles, in Paris ; British Museum, *Catalogue of the Greek and Etruscan Vases in the British Museum ;* Berlin, catalogue by Furtwängler, 1885 ; Munich, Otto Jahn, *Beschreibung der Vasensammlung König Ludwig's in der Pinakothek zu München*, 1854 ; Naples, Bourbon Museum, Heydemann, *Vasensammlung des Museo Nazionale zu Neapel,* 1874 ; St. Petersburg, Museum of the Hermitage, catalogue by Stephani, 1869 ; Athens, Museum of Varvakeion : Collignon, *Catalogue des Vases peints du Musée de la Société arch. d'Athènes,* 1878.

To Winckelmann belongs the honour of discovering the Hellenic origin of painted vases; and James Millingen in the beginning of the present century founded a school of simpler and more reasonable criticism. The discovery of the necropolis of Vulci in Etruria, near the Ponte della Badia (1828), is a most important item in the history of the study of painted and inscribed vases. Many thousand painted vases were brought to light, and the work of Gerhard on the vases of Vulci marked the beginning of a new school of research, which substituted knowledge for guesses. French and other foreign scholars, among whom Gerhard, Panofka, Otto Jahn, Charles Lenormant, and De Witte, occupy the front rank, have introduced method into these studies.*

While at the present time an exclusively Etruscan origin is no longer attributed to painted vases, the question as to the place of their production remains one of the most difficult of archæological problems. From whence came those vases of Greek type found in Etrurian tomb-cities, styled Italo-Greek vases? Were they imported from Greece, or is it better to suppose them the products of local workshops, imitations made by Italian artists of models brought from Greece? We cannot admit the opinion of Otto Jahn, advocated previously by G. Kramer, according to which all vases found in Italy, with a few rare exceptions, were produced in Greece, perhaps even in Athens. This rigid system leaves no place for the

---

* Furtwängler, whose catalogue of the vases in the Berlin Museum is a model, promises an *Handbuch über die griech. Vasenmalerei.*

activity of Italian potters.   The solution of this question can only be attained by the careful comparison of Italo-Greek vases with those of Greece proper.

There can be no doubt that in the sixth and fifth centuries B.C. importations from Greece into Italy were very frequent : this is proved by fact.   In Greece and in Italy have been found vases identical in style, with similar subjects ; and this coincidence is especially striking in the archaic period.   The testimony of history confirms that of these facts.   It is known that about 655 B.C. (Olym. XXXI. 2) the Corinthian Demaratus, driven from Corinth by the tyrant Cypselus, emigrated to Tarquinii, accompanied by two artists, Eucheir and Eugrammus.*   The legendary name Eugrammus, "one who traces beautiful lines," seems to signify simply that Demaratus brought with him potters skilled in the art of decorating vases. Thus does the existence in Etruria of Greek earthenware, covered with inscriptions in Corinthian characters, find an easy explanation.   The abundance of the commerce between Greece and Italy in later epochs is furthermore proved by definite facts.   Certain signatures of artists are to be read at once on vases found in Greece, and on pottery from the tomb-cities of Italy.†   This argument settles the question.

---

* Pliny, *N.H.* xxxv. 152.

† A. Dumont, *Peintures céramiques de la Grèce propre*, p. 5. The names recovered by Dumont on vases from Greece proper are the following : (1) on black-figured vases, Chares, Cheiron, Ergotimus, Exekias, Nearchus, Scythes, Timonidas, Tleson, son of

It is beyond doubt, however, that at one time the Italo-Greek districts had their local workshops, and that Greek vases were imitated in Italy. We draw this inference from the study of Italian pottery, which betrays, especially at the time of decline, a very marked local style. To sum up, we must admit frequent importations in the sixth and fifth centuries B.C., community in inspiration, imitation in Italy of types peculiar to Greece proper, commercial relations at all times, and also the importance of local fabrics, which increases the further we are separated from the best epoch. Such seems to be the truth on the question as to the resemblance between Greek and Italian pottery.*

We shall, then, be justified, in our general discussion, in applying the same principles to Italo-Greek vases as to vases from Greece proper, and to regard them all as products of Hellenic industry which wrought under the influence of the same traditions in all the Greek countries bordering on the Mediterranean.†

Nearchus, Pasias, Chelis, Nicosthenes, Gamedes; (2) on red-figured vases, Hegias, Hilinus, Psiax, Xenophantus. The names of Cachrylion, Teisias, Procles, and Megacles, should be added to the list. The signatures of Greek artists upon vases have been collected by Klein, *Die griech. Vasen mit Meistersignaturen* (*Denkschriften der Wiener Akad.*), 1883.

* Dumont, *id.*, p. 25.

† It should, meantime, be noted that though painted vases are Greek and not Etruscan, Etruria had a special industry—that of vases of black clay with low reliefs, the so-called *bucchero nero*. These are found usually in a limited territory, between the Tiber on the south, and Sienna on the north. This is the national pottery of Etruria. [Specimens have been found in Rhodes, etc.]

CHAPTER II.

THE FORMS AND TECHNIQUE OF PAINTED VASES.

§ I. FORMS OF VASES.

THE nomenclature of Greek vases is far from being definitely established. Among the varied forms created by the fancy of the potter there are some that we cannot designate by special terms. Thus, in collections of Greek vases, it is customary to direct the reader, in the case of each vase, to a table of general forms, in which the forms are known by numbers. This method, while scientifically precise, has the advantage of supplying means of reference to forms for which no name is given. And yet there are certain designations that are clearly established and definite. The important work of Panofka,* revised and completed by other scholars, has fixed the meaning of a great many names. These names have been adopted by all archæologists of the present time, and apply to

---

* Panofka, *Recherches sur les véritables Noms des Vases grecs,* 1829; Letronne, *Observations sur les Noms des Vases grecs,* etc. Paris, 1833; *Cf.* Ussing, *De Nominibus Vasorum,* 1844 ; Ch. Lenormant and de Witte, Introduction to their *Élite des Monuments céramographiques;* Lau, *Die griechischen Vasen, ihre Formen und Decorations System,* 1877

well-recognised forms. We will here mention the most common and at the same time the simplest types.

The *amphora* (ἀμφορεύς) is a vase with ovoidal body, mounted on a foot that sometimes has the form of a truncated cone. Two handles are attached to the neck and to the upper part of the body. The dimensions of the amphora vary from the amphoridion used for domestic purposes up to the magnificent and richly painted amphora which figured among wedding presents in a nuptial procession, as may be seen on a beautiful vase in the Museum at Athens. These great amphoræ were simply objects of luxury, since specimens of them found in our museums seem to have no hollow interior, and could not, therefore, have been put to any use.

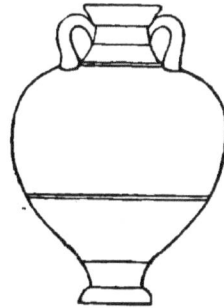

FIG. 84.—AMPHORA.

The *crater* (κρατήρ) was generally of great size. According to the etymology of the name, it is the vase used for mixing water and wine. Its form is spreading, the mouth or opening is wide, and two small handles are attached to the lower part of the body. The *oxybaphon* is derived directly from it, with this difference, however—that the handles are placed higher up, and are attached just under the orifice of the vase. The orifice itself, instead of expanding like the calyx of a flower, as in the crater, is joined to the body by an oblique moulding.

The *celebe* (κελέβη) presents forms similar to the

two preceding, but at the neck it grows narrower,

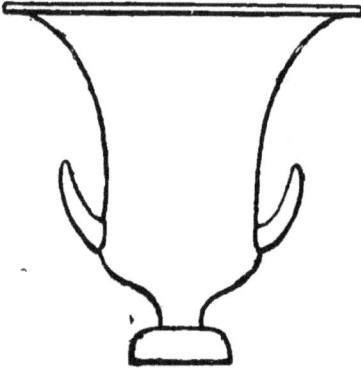

FIG. 85.
CRATER.

FIG. 86.
CRATER: HANDLES WITH VOLUTES.

FIG. 87.—OXYBAPHON.

FIG. 88.—CELEBE.

and the two handles are fixed into the lower surface

of a large flat border which crowns the opening of

FIG. 89.—STAMNUS.

FIG. 90.—HYDRIA.

the vase. The *stamnus* (στάμνος) is likewise narrower at the neck, and has two handles or ears placed upon the upper part of the body of the vase.

The *hydria* (ὑδρία) is characterised by the neck that surmounts its ovoidal body, and by its three handles, one of which, at the back, is attached to the mouth and rises above it ; the other two, like elbows, are applied to the sides of the vase. The *calpis* (κάλπις)

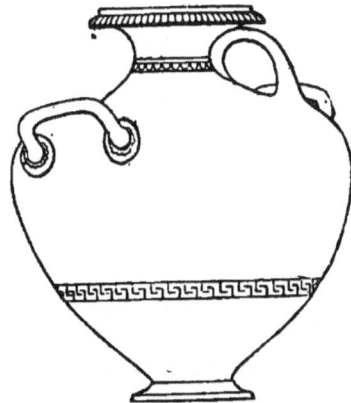

FIG. 91.—CALPIS.

differs from the *hydria* only as to the handle at the

back, which is lower ; the *pelice* (πελίκη) has only two handles placed at the neck of the vase, which is joined to the body by a curve scarcely perceptible.

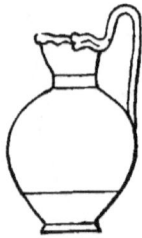

FIG. 92.
ŒNOCHOË.

The *œnochoë* (οἰνοχόη) is usually of smaller dimensions, and exhibits most elegant outlines. The single handle which adorns the back is delicate and is gracefully curved ; the edges of the orifices at three places are, as it were, bent inward, and thus outline a trefoil. Not uncommonly, at the best period of ceramic art, the œnochoë, like other vases of small size, is decorated with most exquisite paintings; certain speci-mens from Attica are perfect masterpieces. The *prochoös* (πρόχοος) differs but slightly from the œnochoë. It is the vase that the Homeric poet puts into the

FIG. 93.—CYLIX.

hands of cup-bearers in the *Oayssey*.* The *epiclysis* (ἐπίχυσις) is also a variety of the œnochoë.

Amongst the most elegant vases should be placed the cup or *cylix* (κύλιξ). Sometimes it is without a

* Hom., *Odyss.* xviii. 397.

foot (*apodal*) ; sometimes it rests upon a foot, light
in design, above which it expands to a considerable
size. The cylix is more or less shallow ; frequently
it is almost flat. The *cyathus* (κύαθος) is a cylix with
a single handle, and the *holmus* (ὅλμος) is a sort of
cylix with a very slender foot, and is without a handle.

FIG. 94.—CYATHUS.　　　　　FIG. 95.—CANTHARUS.

When the flat handles of the cylix are attached to a
vase without a foot, gradually diminishing in diameter
towards the base, the vase is called a *scyphus* (σκύφος).

The *cantharus* (κάνθαρος) is pre-eminently the
Dionysiac vase ; * it is a large cup, furnished with two
very elevated handles ; it has a foot, and was often

---

* In the combat between the Centaurs and the Lapithæ, painted by
Hippeus, the contestants strike at each other with canthari. *Cf.* Athenæus, xi. 474 D, on the carchesion. [In Athenæus, *Deipn.* xi., are given
the names of many Greek vases and cups, with descriptions.]

decorated with subjects taken from the cycle of
Dionysus. The *carchesion* (καρχήσιον) differs from it

FIG. 96.
CARCHESION.

FIG. 97.
LECYTHUS.

FIG. 98.
OLPE.

in its dimensions, and in its handles, which are
attached by a sort of bolt to the
sides of the vase.

FIG. 99.— ARYBALLIC
LECYTHUS.

The carefully moulded and
delicate forms of certain other
vases, as well as their small size,
indicate that they were designed to
receive liquids more precious than
wine or water. Such is the *lecythus*
(λήκυθος), a sort of cruet used for
perfumes, where the slender body
is terminated by an elegant neck,
with a conical opening; this type,
when of Athenian workmanship,
often presents forms of rare beauty. The *olpe* (ὄλπη),
derived from it has a more rounded body, and a more

open neck. The handle is very elevated, and is attached at the edge of the orifice and at the base of the neck. When the lecythus, instead of being elongated, becomes rounded and has a spherical body, it is called an *aryballic lecythus;* it resembles the *aryballus* (ἀρύβαλλος), which is a spherical vase without a foot with a contracted neck, which terminates in a wide flat brim, and is attached to the body of the vase by a very short handle. The aryballus was used to hold

| FIG. 100. | FIG. 101. | FIG. 102. | FIG. 103. |
| ARYBALLUS. | BOMBYLIUS. | COTYLISCUS. | ALABASTRON. |

the oil with which athletes rubbed themselves down; it figures often in bathing scenes, and in scenes of the palæstra and of the gymnasium. The *bombylius* (βομβύλιος) is an elongated aryballus; it has a flat brim, furnished with a solid handle pierced with a small hole. With a body very slender at the base, but widening above, it becomes the *cotyliscus* (κοτυλίσκος).

These vases are not all designed to hold liquids. The *pyxis* (πυξίς) is in fact a toilette box, having a body which is set into a lid, furnished with a ring of bronze. Scenes from the toilette, which often ornament the

S

lid, recall the uses of the pyxis; in one of these vases, discovered in Athens, were found pastilles of paint. The *alabastron* likewise had its place in the toilette. In the scenes represented, it is found in the hands of goddesses, or of attendants attiring their mistresses. It is a vase of elongated form, with a narrow neck, and is sometimes of alabaster or of coloured glass.

FIG. 104.— RHYTON.

In this rapid survey we are far from having exhausted the series of forms created by Greek potters. One can obtain an idea of the fertility of their invention only by seeing on the shelves of a museum these varied shapes, often borrowed from types of animal and of vegetable life. A vase now represents a hare, or a bird; now it is a human foot sandalled; again it represents two shells fitted together. In this class of vases the most remarkable is the *rhyton* (ῥυτόν), which often has the form of a curved horn; the pointed part represents the head of an animal, of an ox or of a horse, surmounted by a large spreading neck, to which a handle is attached. At times the rhyton is no more than a drinking-horn, from which may flow a thin stream of liquid; it is then an actual receptacle, furnished with a foot, the central part being formed by a head in relief. Such is a beautiful rhyton in the Museum at Athens, representing the head of an Ethiopian, with lips painted a vivid red. When potters abandoned classic forms their fancy knew no bounds. The vase was decorated with reliefs, which became the essential part. These mixed products belong as much to the class of terra-cottas as to that of painted vases.

§ 2. TECHNIQUE OF VASES.

It is chiefly to the labours and experiments of the Duc de Luynes* that we are indebted for our knowledge of the manufacture of Greek vases. The minute analyses of this scholar have thrown light upon the principal points in question.

The clay used by the potters was very fine, and was carefully prepared ; the vase was made on the potter's wheel, and the workman afterwards attached the neck and handles. After the first baking, which left the clay still soft, the artist, whose business it was to paint the vase, traced his subject upon it with a blunt or rounded point, marking the main outlines.†

The process of painting varied, according as the vase was decorated with black figures on a red ground (black-figured vases), or with red figures on a black ground (red-figured vases).

In the former case it is the natural colour of the clay that gives the red ground for the painting. The figures stand out like black silhouettes, which the artist obtains by filling in the outlines of the sketch with

---

* *Annali dell' Instituto di Corr. arch.*, Vol. IV., p. 138 *ff.*

† We can still see upon certain vases traces of this sketch, as on a cylix in the Museum at Athens : Collignon, *Catal. des Vases peints du Musée d'Athènes*, No. 462 ; the same on a cylix signed Cachrylion, *Bull. de la Société des Antiquaires de France*, 1878, p. 47. These tracings constituted the first sketch, by means of which the artist essayed to define his subject.

S 2

colour ;* details, such as the muscles, the folds of garments, the features of the face, etc., are drawn afterwards with a dry point, which attacks the black tint, and causes the natural colour of the clay to reappear. This is the most ancient method. It is followed in vases of the archaic style.

When vases were ornamented with red figures on black ground, the technique was very different. The sketch was made with a fine brush charged with black paint ; the artist with a heavier brush then surrounded these outlines with a bold uniform tint which isolated them, and afterwards retouched with a black coating all the ground of the vase. The details of the figures, thus standing free upon the red colour of the clay, were then traced with a fine pencil in lines of exceeding delicacy.

Black was not the only colour employed ; for retouching, use was made of white, and of purplish-red, to heighten the details of the black figures. Later, polychrome paintings became popular, especially in the fourth and third centuries B.C. To the most carefully finished vases were applied gildings that heightened still more the richness of the painting, upon which blue, green, bright yellow, and brownish-red were lavished. Finally, certain vases, particularly the lecythi of Athenian manufacture, were covered with a white coating, which was carefully polished, and easily lent itself to work with the brush.

Inscriptions accompanying the figures, or relating

---

* This black colour had for a base oxide of iron.

to the artist, are traced with a brush, either with the brilliant black used for the figures, or with the other colours used in retouching ; sometimes inscriptions are engraved with a sharp-pointed instrument after the manner of *graffiti*.

The work in the vases is thus twofold : that of the potter, and that of the designer and painter. This co-operation is attested by the signatures of artists placed upon vases.* Thus the François Vase of the gallery of Florence, one of the most beautiful black-figured vases known, bears the double signature of the potter, Ergotimus, and of the draughtsman, Clitias, Ἐργότιμος ἐποίησεν, Κλιτίας μ' ἔγραψεν.† The word ἐποίησεν usually refers to the modeller, who gave to the vase its elegant form, and the word ἔγραψεν to the artist who covered it with figures, which were first drawn and then painted.‡ Sometimes the potter is also the designer. In that case his name is usually followed by the phrase, ἔγραψεν καὶ ἐποίησεν. These signatures are most important in the study of ceramic art.

It is not easy to define the amount of invention that belongs to the designer in the decoration of vases ; it is certain that he employed neither pointers nor tracings. Sketches in which we can follow the outlines

---

* See De Witte, *Noms des Fabricants et Dessinateurs des Vases peints*, 1848.

† ΕΡΓΟΤΙΜΟΣ ΕΠΟΙΕΣΕΝΕΝ *(sic)* ΚΛΙΤΙΑΣ ΜΕΓΡΑΦΣΕΝ.

‡ The word γράφειν refers not to the γραφαί, paintings, but to the γράμματα, which are the lines drawn. It was here that the talent of the artist showed itself. An ordinary workman could fill in with colours the outlines sketched by the more skilled artist.

indicate the gropings, as it were, of the artist, who is feeling his way. But did he copy a model? We admit that the pictures of the most famous painters might have been reproduced by the painter of vases ; it is exceedingly probable that for certain common subjects the artist had a model before his eyes, which he imitated more or less exactly, reducing or adding figures, according to the space to be filled up. But he was not obliged to make a slavish copy ; numerous vases certify that originality and fancy asserted themselves. The artists in vases were assuredly of an humble order. And yet in a people, the most artistic that has ever existed, the humblest works recall the traditions of the grand style. Art and trade were not strictly divided, and in his limited domain the painter of vases could attain and maintain a certain independent individuality.

CHAPTER III

CLASSIFICATION OF PAINTED VASES.

THE safest method in the study of ceramic paintings consists in classifying them in chronological order according to the style of their decoration. In this way groups are formed which present distinct characteristics, and correspond to the different ages of ceramic industry. No one can fail to recognise the importance of this classification in the history of Greek art ; it is through the succession of styles that we can follow, from one group to another, the development of taste in general, which left its mark on painted vases, as well as on the most beautiful marbles. It should be added that of this class of monuments the most numerous specimens are preserved ; it thus offers a continuous series from the beginnings of Greek art to the time of decline.

Painted vases range themselves at once into three principal groups, each of which has several subdivisions :—

1. VASES OF ANTIQUE STYLE. .

2. BLACK-FIGURED VASES.

3. RED-FIGURED VASES, AND VASES OF LATER STYLE.

We will endeavour to state briefly the characteristics in technique of the special varieties under each of these groups.*

§ 1. VASES OF THE ANTIQUE STYLE.

(1) *Santorin pottery.*—It is well known that the vases found at Santorin under the pozzuolana, and earlier in origin than the submergence of the island, are among the most ancient remains of civilisation in Hellenic countries.† They go back as far as eighteen or twenty centuries B.C. The ornaments that decorate them are very simple, and more commonly are taken from the vegetable kingdom. One characteristic feature of this barbaric pottery is the imitation of the human form, and in particular of the female head and bust, which the potter has sought to reproduce in modelling in clay.

(2) *Vases of the antique style of the Cyclades.*—The designation *Phœnician vases of the Cyclades*, sometimes applied to these vases, is far from being accurate ; it is admitted, however, that the date of the manufacture of this pottery coincides with the Phœnician sway in the Greek Archipelago, and is not later than the twelfth or thirteenth century B.C. These vases are found in the Greek islands, Milo, Santorin

---

* See A. Dumont, *Peintures céramiques de la Grèce propre.*

† See Book I., chap. i. The Museum of the École Française in Athens possesses a rich collection of unpublished vases. Several are reproduced in the work by MM. Dumont and Chaplain, *Les Céramiques de la Grèce propre,* plates I. II.

(Thera),* Rhodes, and Cyprus. These vases are often

FIG. 105.—VASE OF THE PRIMITIVE STYLE OF THE CYCLADES.

* The vases of this style from Santorin are subsequent in origin to the submergence of the island, and have been found above the pozzuolana.

great jars, with a grey earth-coloured ground, decorated in zones, curved lines, zigzag lines drawn in dull brown sometimes heightened by a retouching of lilac or light pink. The human figure no longer appears. The specimen here given (Fig. 105) is from Thera ; the ground of the clay is whitish, and its decoration, consisting of concentric circles, of zones, and of chevrons, is in dark brown or orange.

(3) *Vases with geometric decoration.*—These vases are found all over Greece, in Mycenæ, in Ægina, in Attica, etc. They are the product of a national art, that owed nothing to foreign imitations, and for this reason deserve serious examination.* These vases differ from the type of the Cyclades in their regular system of ornamentation, which is essentially rectilinear or geometric. The most important group of this class is formed by pottery found at Athens. It is to these vases that we are indebted for the detail of the ornamentation.†

The forms of the vases vary from the amphora down to the smallest cups. The paintings are done in red-brown, passing sometimes into black, upon the reddish ground of the clay. The decoration consists of meander patterns, oblique lines, chevrons, rosettes, concentric circles, often most carefully executed ; on

---

* See A. Conze, *Zur Geschichte der Anfänge der griechischen Kunst,* 1870—1873.

† See Hirschfeld, *Annali dell' Inst.,* 1872. The pottery of Mycenæ forms the subject of a special work, of which a portion has already appeared : *Mykenische Thongefässe,* by Furtwängler and Lœschcke, Berlin, 1879.

some vases the rosettes are outlined with compasses.
It is quite probable that this pottery was nothing
more than an attempt to reproduce the decoration of
metallic vases. This hypothesis is all the more
plausible in that certain clay tripods betray in their
forms and in their modelling an evident imitation of
metallic tripods.* To these geometric patterns were

FIG. 106.—GEOMETRIC STYLE, FROM MYCENÆ.
(From Dr. Schliemann's *Mycenæ*.)

often added representations of animals : horses, stags,
deer, and birds. These figures of animals, executed
rudely and awkwardly, have a peculiar type of their
own, which prevents them from being confounded with
those of the following period, where Oriental imitations
are very clear. The human figures, which are arranged
in zones on Attic vases of geometric style, have all

* Conze, *l.c.*, plate VII.; Collignon, *Catal. des Vases peints d'Athènes*,
No. 31.

the characteristics of the most primitive art ; they are
no more than very rude silhouettes. The breast is
disproportionately large, the waist slender, the thighs
have an exaggerated development. The scenes
represented are processions, warriors on their chariots,
funeral obsequies, and the laying-out of the dead
(πρόθεσις).* The plan of this brief sketch makes
it impossible for us to enter into the theories to
which the peculiar characteristics of this geometric
style have given rise. It is sufficient to recall the fact
that this style is found in northern countries, and that
it must have been common to the Indo-European
people before it broke up into separate branches.
With regard to pottery of this kind from Greece
it must be admitted that no date later than the tenth
century B.C. can be attributed to it. Whether, with
Conze, we give these vases the name *Pelasgic*, or a
totally different name, it is beyond doubt that they
were manufactured in Greece throughout a long
period, before relations with the Asiatic Orient had
furnished Greek potters with the models which afterwards became the inspiration of ceramic art.†

(4) *Vases from Milo.*—Oriental influences are
clearly apparent upon vases found at Milo that
date from the eighth or from the seventh century B.C.

---

* The most remarkable specimens are in the office of the Minister
of Education in Athens. [Fig. 107 represents the πρόθεσις. Certain
vases of this style are also called Dipylon vases, from the locality
(Dipylon at Athens) where they were found.]

† This style was preserved much later, probably through the influence of tradition.

By the side of rectilinear decorations, which are sur-
vivals of the ancient geometric style, are seen zones
of animals Oriental in character, and decorative motives peculiar to Asia, such as chimæras face to face.
At the same time, Greek gods are figured in their Hellenic forms.   These vases form a transition series,

FIG. 107.—FRAGMENT OF AN ATHENIAN VASE OF ANTIQUE STYLE.

and are nearly contemporary with the earliest vases
of the following class.*

(5) *Vases of the Corinthian* or *Asiatic style.*—These
vases are often called Corinthian vases, because a
number of them have been found in tombs in the
neighbourhood of Corinth.   They are found, however, in all portions of the Hellenic world, and even
in Etrurian tomb-cities.   The distinctive characteristic of these vases is their decoration, the motives of
which come immediately from the East.   On them

* See Conze, *Melische Thongefässe,* 1862.

we find the rosette of Assyrian monuments and figures of fantastic beings, half-human, half-animal; birds with human heads wearing the Oriental polus; flying personages with wings curved backward—all of them symbols which had significance only in the East, and which were copied by the Greeks without being understood. The most frequent forms of this class of vases are the spherical aryballus, the bombylius, the alabastron, and the deep cylix. The clay is yellowish-white, and the paintings, according to the period, are dull or more vivid, and finally even of an intense black relieved with purple and red.* The manufacture of these vases continued for quite a long time, up to the period of black-figured vases

FIG. 108.
BOMBYLIUS FROM TANAGRA,
CORINTHIAN STYLE.

* The Louvre possesses a rich collection of them. Specimens may be found in the work of M. de Longpérier, *Musée Napoléon III.*

properly so called. As it is very difficult to class vases chronologically, we must limit ourselves to grouping them according to subjects in the following order :—1, vases with zones of animals ; 2, vases with human figures ;
3, vases with mythological subjects and inscriptions.

1. These vases, which are of large size, are decorated with several zones of animals, such as lions, goats, tigers, antelopes, usually figured as fronting each other, sometimes as marching in file. The colours are often retouched in purplish-red, while the details of the muscles are indicated by lines drawn with a dry point. Rosettes fill up the field of the zones.

FIG. 109.—CORINTHIAN VASE WITH ZONES OF ANIMALS.

2. Upon vases with figures of persons, subjects taken from Greek mythology are represented between zones of animals. This decorative system is borrowed directly from the East. The Greeks copied it

either from stuffs and carpets woven in the East, or from metal cups of Cyprus or of Assyria, through the instrumentality of the Phœnicians.

3. At length, in the seventh century B.C., inscriptions in Corinthian characters appear upon vases with mythological subjects. Persons are designated by their names, traced in archaic Greek letters, which are those of the Corinthian alphabet in the seventh century B.C. The most remarkable specimen is a pyxis found at Mertese (Dodwell vase), which represents the Calydonian Boar Hunt. Each person has his name : Θερσανδρος, Φιλον, Λακον, Ανδρυτας, Σακις, Αλκα, Δοριμαχος, Αγαμεμνον. It is upon vases of this series that the earliest signatures of artists may be read, as that of Chares, or that of Timonidas of Corinth on a vase in the museum at Athens representing Achilles watching for Troïlus.

The manufacture of this class of vases was certainly very widely extended. Commerce scattered them in all parts of the Hellenic world ; they are to be found in Etruria, and the presence in this country of the Corinthian colony led by Demaratus must have contributed to bring this style into favour. From Cervetri—ancient Agylla (or Cære) of the Etruscans—come the greater part of the vases of the Corinthian style now in the Louvre :* they form an exceedingly rich collection, without its equal in Europe. We shall mention only one of the best known vases from this collection, a large celebe,

---

* The Campana Collection.

decorated with a band of animals, and above with a half-zone of figures. The scene represented is taken from Homer, and represents the departure of Hector,

FIG. 110.—CORINTHIAN VASE, WITH FIGURES OF PERSONS AND INSCRIPTIONS.—DEPARTURE OF HECTOR.

who is about to mount his chariot. The persons that surround the hero are Trojan warriors and members of the family of Priam. (Fig. 110.)

There is a general agreement in dating the earliest of these vases with inscriptions from the earlier half of

T

the seventh century B.C. (660 B.C.). The date is in this case of great importance, as it enables us to establish the interesting relations existing between the style of vase-paintings and that of other contemporary monuments. We have already spoken of the celebrated chest of Cypselus,* that was for its time one of the wonders of metal-work ; the style of the figures with which it was covered, the arrangement of the subjects upon it, its inscriptions, written *boustrophedon*,† *i.e.*, from left to right and then from right to left alternately—all these items are explained by an examination of Corinthian vases. Admitting for the chest of Cypselus the date of the thirtieth Olympiad (660 B.C.), we see that it is older by only a few years than the most ancient Corinthian vases. It cannot be doubted that these monuments exhibit the same inspirations, the same methods. This was the age when Greek genius had hardly detached itself from Asiatic influences. By the side of subjects purely Hellenic are to be found foreign motives of decoration borrowed from the East.

The manufacture of Corinthian vases reaches back to the period of primitive vases. A large number of these ceramic products also exhibit the technical methods pursued in the following period, and nearly approach the black - figured ware. Thus upon Corinthian vases of the later type the flesh of female figures is painted in white, as are also the long tunic

---

* See Book I., ch. iii., p. 29.

† Βουστροφηδόν, literally, like the furrows drawn by oxen, that return on their steps at the end of each furrow.

of the charioteers, and the *episema*, or emblems, on the shields.

A more detailed system of classification would make a place at the beginning of the following class for those vases in which the principles of black-figure painting are applied at the same time with Asiatic ornamentation. We must here confine ourselves, in order not unduly to multiply divisions, to specifying the class of vases that formed the transition between the Corinthian style and that of black-figured vases.

§ 2.—BLACK-FIGURED VASES.

The period of black-figured vases extends from the sixtieth to the eightieth Olympiad (540 B.C.— 460 B.C.), not to mention ceramic products of a later age, in which, even until the fourth century B.C., the same process was revived and imitated. The style of these paintings, like that of the primitive masters, has all the marks of archaic art: stiff figures, almost always represented in profile, angular action, faces without expression, and uniform in type. We may add that even after the decline of painting in black these characteristics were preserved in a conventional way. We then find intentional faults, and awkwardnesses too skilfully represented to be genuine. Thus Panathenaïc amphoræ, some of which are of the time of Alexander, exhibit all the features of archaic paintings, although they are contemporary with the most beautiful red-figured vases. Here archaism is affected, and is simply conventionality.

T 2

We have already indicated the technique of painting
in black, the use of white colour to distinguish
women from men, and that of purplish-red to heighten
the effect of the paintings. It is unnecessary to
enlarge further upon this point.

The subjects most commonly figured on this class of
vases are taken from mythology, and principally from
the Dionysiac cycle. Processions of gods are very
frequent, and these vases are of great interest in the
study of the plastic types of Hellenic divinities.
Scenes from the Trojan war, the Labours of Heracles,
and Attic myths, especially the myths of Theseus,
furnished these vase-painters with most of their
subjects.

A compendious classification enables us to divide
black-figured vases into several series, the principal
of which are the following : (1) vases with white
or yellow ground ; (2) vases of the style of Ergotimus
and of Clitias ; (3) vases of the style of Nicosthenes ;
(4) vases of the severe style ; (5) Panathenaïc amphoræ ; (6) common products.

(1) *Vases with white* or *yellow ground.*—Vases of
this series, where black figures stand out upon a coating of yellowish white, are still very rare. The
Cabinet des Médailles, at Paris, possesses an interesting specimen, known as the Cup of Arcesilas.
The paintings depict this person (probably a king of
Cyrenaïca) * seated under a canopy, surrounded by
attendants who are engaged in weighing in balances

---

* Perhaps the victor at the Pythian games in the seventy-eighth
Olympiad (466 B.C.), whom Pindar celebrates in *Pyth.* iv. and v.

silphium, a precious product of Cyrenaïca, much in demand in Greece.* To the same series belongs another cup in the Cabinet des Médailles, representing Ulysses and his companions intoxicating the Cyclops Polyphemus, and putting out his single eye with a stake heated to a glow in the fire.

(2) *Vases of the style of Ergotimus and of Clitias.* —The signature of these two artists† is read with certainty only on a single vase, known as the François Vase of the Museum of Florence. The style of painting upon this vase is, nevertheless, of sufficient importance to mark one of the periods in the history of Greek ceramic painting. The vase in question is one of the wonders of the art of black-figured vases. It is a magnificent amphora, and is decorated on the body with three zones of subjects. Two other zones ornament the neck and the foot ; even the handles are covered with paintings. The principal subject is a procession of divinities, present at the marriage of Thetis and Peleus, in seven quadrigæ. Other subjects are the funeral of Patroclus, Achilles pursuing Troïlus, the combat between the Lapithæ and the Centaurs, the Calydonian Boar Hunt, Theseus and Ariadne, etc. All these compositions are worked out with singular richness, and are treated in that severe archaic style which lends such a charm to the works of old Greek masters. This vase, which is of rare beauty, makes us under-

---

* Figured, with colours, in Birch, *History*, etc. It can hardly be regarded as a caricature. On these vases, *cf.* Puchstein, *Arch.Zeitung*, 1881.

† Eucherus, son of Ergotimus, is also known by his signature.

stand with what fertility of resource the metal-workers of the Dorian schools had decorated famous monuments, such as the throne of Amyclæan Apollo, known only by descriptions of Greek writers.

(3) *Vases of the style of Nicosthenes.* — The vases of this style are easily recognised by the firmer drawing, but principally by the very characteristic palmette ornament on the neck of the vase. There is in them an elegant combination of lotus flowers and of knots that seems peculiar to Nicosthenes and his school. The accompanying cut shows a vase from the Louvre, signed by this artist. It is decorated with two zones of subjects, one of which represents the Sphinx, and perhaps Œdipus. (Fig. 111.)

FIG. 111.—VASE OF THE STYLE OF NICOSTHENES.

The vases signed by Nicosthenes are numerous. They have been found at Vulci, at Cære, at Agrigentum, and in Attica. It is clear that vases manufactured by him were very celebrated, and the special taste that characterises his works made

them highly prized in different parts of the Hellenic
world.

FIG. 112.—BLACK-FIGURED HYDRIA.
(Style of Timagoras.)

(4) *Vases of the severe style.*—By this name we
designate paintings in which colours are more
rare in the retouching, the effect being chiefly

obtained by engraving with a dry point, which brings
out the details of the body and of costume.   The figures
are drawn with great energy, and with an evident
desire to accent their anatomy.   At the same time,
all trace of Oriental ornamentation has disappeared.
We have now arrived at the second archaic period.
The most beautiful specimens of this class are contemporaneous with works that in sculpture immediately preceded the period of most perfect art.   The
names of artists most frequently found on vases of
the severe style are those of Timagoras, Amasis,
Tleson son of Nearchus, and Hermogenes.

The Louvre possesses a beautiful series of these
vases, from which we select for special mention the
hydria of Timagoras, which exhibits the struggle of
Heracles with Triton or with Nereus. (Fig. 112.) The
severe style appears with all its characteristics in a
painting now in the Museum of Athens.   It is a dish
found at Phalerum, with a painting representing the
Arming of Achilles.   Thetis carries to her son the
divine armour, which the hero is putting on, in the
presence of Neoptolemus and Peleus.*

No other museum in Europe possesses a work
to be compared with the magnificent amphora, decorated with black paintings, found at Cape Kolias,
now in the Museum of Athens (Varvakeion).   The
principal subject is the laying-out of a corpse
(πρόθεσις). Around the bed women weep with gestures
of grief, and seem to accompany these with exclama-

---

* Collignon, *Catal. du Musée d'Athènes*, No. 231.

tions. The scene has a striking effect, which is due to the austere simplicity of the composition, and the expression of sorrow given to the faces. The qualities of the style are those of an epoch that has attained perfection, and black - figured vase-painting has produced nothing more finished than this amphora.

(5) *Panathenaïc amphoræ.* — These vases, which were given as prizes to victors in the Panathenaïc festival, have marked peculiarities of their own. They are in the form of amphoræ, with covers. The painting upon the front of the body represents Athena fully armed, poising her lance as if in battle. On each side of the goddess is a column, which is surmounted

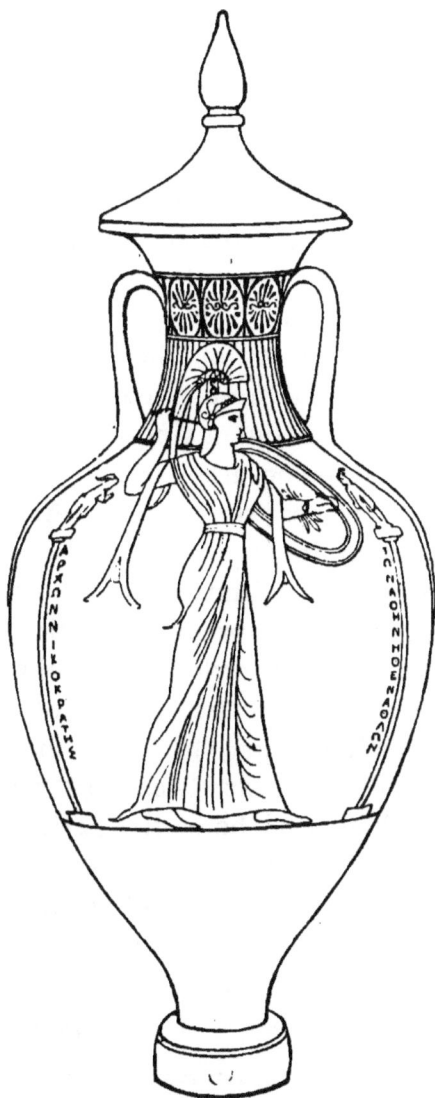

FIG. 113.—PANATHENAÏC AMPHORA.

by a cock, an owl, a vase, or the figure of a person. In the field, and alongside of the columns, are to be read the painted inscriptions, one of which records the purpose of the vase, *ΤΩΝ ΑΘΗΝΗΘΕΝ ΑΘΛΩΝ* (" prize from the contests at Athens"*). The other inscription indicates the archon eponymous in office at the time of the contest. The painting on the back of the vase always indicates the variety of contest for which the prize was awarded. These vases are found in museums in large numbers, and were exhumed in various parts of the Hellenic world, such as Italy, Cyrenaica, and elsewhere. Until recently Athens furnished but a single example. We now recognise Panathenaïc amphoræ in several fragments found in Athens near the temple of Athena Polias. These vases had undoubtedly been consecrated to the goddess. It is also admitted that for this class of vases black figures are not an indication of antiquity. Though the archaic style is still followed, it is only through tradition, for the names of archons in office in the fourth century B.C. may be read upon them, at a time when art was far removed from the naïve archaism of the earlier period.

(6) *Common products.*—We place under this heading a series of vases of which great numbers are found in museums, especially in Athens. They have been but imperfectly studied. The most numerous are lecythi, decorated with paintings rapidly and often

---

* On the oldest vases the inscription has the peculiarities of the ancient Attic orthography ; *i.e.*, ΤΟΝΑΘΕΝΕΘΕΝΑΘΛΟΝ, τὸν 'Αθένεθεν ἄθλον.

carelessly executed. Several types are to be distinguished in this series of vases: 1. Those of the
type of Phalerum and Bœotia, recognisable by their
awkward forms, and by the yellow tint of the clay.
Lecythi of Phalerum have upon their necks a cock
between two ivy leaves. Ephebic and Dionysiac scenes
are frequently figured. 2. Athenian type; the forms
are more elegant, and Attic myths furnish the principal subjects for decoration. 3. Locrian type; the
figures are drawn in black upon a yellowish ground.
One of the most interesting specimens in the Museum
at Athens represents Dionysus punishing the Tyrrhenian pirates, who are metamorphosed into dolphins.
This subject is also found on the frieze of the
choragic monument of Lysicrates in Athens. (See
ante, p. 95.)

§ 3. RED-FIGURED VASES AND VASES OF LATER STYLE.

Vases of this class are by far the most numerous,
and the varieties of style, which correspond to the
advance and decline of art, are clearly marked. They
have certain general characteristics that are in direct
contrast with those of black-figured vases. Every
trace of conventionality has vanished; artists throw
off methods laid upon them by tradition, and aim at
true and independent expression. Compositions are
less overloaded with figures. In place of belts with a
long row of figures, subjects less ambitious are found,
especially upon vases of the best workmanship. At
the same time the execution is more finished, and the

details of the figures, drapery, and costume, are treated with exquisite purity of taste.

The manufacture of these vases had certainly begun before the close of the preceding period. We know of examples that show the simultaneous employment of the two classes of painting. Besides, red-figured vases have been found beneath the débris of the old Parthenon, burned by the Persians in 480 B.C. (Olympiad LXXV.). But though the oldest vases of this style may have been contemporary with the later black-figured vases, it must be acknowledged that the new style was not slow in supplanting the old, and that the age of red-figured vases coincides with the very long Hellenistic period. Materials are wanting by which precise limits may be fixed. In the view of most scholars the manufacture of these vases had ceased by the first quarter of the second century B.C. The Roman senatus-consultum on the Bacchanalia (186 B.C.), interdicting the ceremonies of the Bacchic cult, would at the same time have put an end to the pottery industry, which furnished the accessories necessary for these festivals.

The vase-paintings of this third period present a great variety. They may be grouped into series, the most important of which are the following :—(1) vases of the severe style ; (2) vases of the second period of red-figured ware ; (3) vases of the Attic style, of great elegance ; (4) common products ; (5) vases enriched with gilded ornaments ; (6) vases with reliefs ; (7) white lecythi of Athens.

(1) *Vases of the severe style.*—De Witte well

characterises the vase-paintings of this period : —
" The compositions of the severe style have somewhat of stiffness. In them are again found, in the
expression of the face, the forms common upon black-
figured vases, but we feel that art is about ready
to break the barriers that obstruct her free course.
Hair and beard arranged and treated with great care,
curled locks, garments with stiff folds falling straight,
characterise the severe style." *   In dating these vases
we cannot go further back than the first years of the
fifth century B.C., nor come down later than the beginning of the fourth century B.C.   This is the epoch of
perfected art, of the great schools of Attica and of
Peloponnesus.   As, however, progress is not made at
the same pace in the domain of the different arts,
vase-painters remained faithful much longer than did
sculptors to a certain stiffness of form which was not
completely given up until the fourth century B.C.   In
spite of these slight differences, it is plain that the vase-
paintings of this epoch reflect the style of the wonderful
fifth century B.C.; in fact, the ceramic artist must often
have received inspiration from the masters of painting.   By means of a magnificent amphora in Munich,
representing the rape of Orithyia by Boreas, Welcker
has established the affinity between the style of this
composition and that of Polygnotus, who flourished
about Olympiad LXXX. (460 B.C.).   The Munich vase
reproduces the imposing manner of the painter who
decorated the Lesche at Delphi.   Museums are very

* De Witte, *Études sur les Vases peints.*

rich in vases of this class, but in the front rank should be placed the beautiful amphora of the Varvakeion in Athens, which depicts a scene of lamentation for the dead, and a procession of knights, making ready, with lowered lances, to form an escort.

Without delaying to multiply examples, we will mention the principal vases bearing the names of artists, especially those that betray an independent manner. Andocides was one of the earliest artists in this period. An amphora signed by him,* of a style still stiff, shows the simultaneous employment of black and red paintings. Epictetus belongs also to this transition period. Although he was at times associated with Nicosthenes and Hischylus, who still painted in the black style, he placed his signature only on red-figured vases finely and carefully executed. Sosias excelled equally in the rendering of details, which he treated with rare power. He is the author of a celebrated cup in the Berlin Museum that shows on the inside the great divinities, and on the outside Achilles caring for the wounded Patroclus. Euphronius, who ordinarily attaches the verb ἐποίησεν to his signature, is often associated with Cachrylion.† His name, however, is read on only one cup. This is a beautiful piece of work of the best style, and represents on the outside the exploits of Theseus, and on the interior the hero together with Athena and Amphitrite.‡ In this composition, which is of a severe elegance, De Witte recognises a more or less direct copy of the paintings

---

* *Bull. de l'Inst. arch.*, 1845.   † *Cf.* Klein, *Euphronios,* 1882.
‡ Edited by De Witte, *Monuments grecs de l'Association, etc.*, No. 1.

executed by Micon for the Theseum of Athens. The

FIG. 114.—RED-FIGURED CRATER.

(Vase of Euphronius.)

vase of Euphronius, reproduced in Fig. 114, is a crater

in the Louvre, representing the contest between Apollo and the giant Tityus. Cachrylion is known by about ten vases, showing compositions drawn gracefully and elegantly, but still impressed with archaic severity, though the attitudes of the figures are more supple and less constrained. We have also in this period the works of Pamphæus, of Duris, and of Brygus. The signature of the latter is upon a cup representing the last night in the history of Homeric Troy, and the massacre of the children of Priam.

(2) *Vases of the second epoch of red-figured ware.*— In the fourth century B.C. vase-painting participated in that evolution in art which brought about a taste for elegance of form. As in painting on a large scale the art of Zeuxis and Parrhasius is removed from the simple religious dignity of Polygnotus, so the style of the ceramic painters, contemporary with the later artists, shows the same studied effort after elegance and grace. Robust, manly forms give place to youthful figures; to stiff attitudes succeed charming poses, simple and natural ; the folds of drapery become more graceful, and float, as it were, about the body, instead of permitting the contours to be seen under their translucent folds, as in the severer style. To this period belong the beautiful amphoræ from Nola, which are the ornament of the Naples Museum, such as the vase depicting the last night of Troy, and the stamnus representing Bacchantes. We mention also, from the Museum of the Louvre, a charming cup, representing the poet Linus (Λίνος) giving a lesson in singing or in reading to young Musæus (Μουσαιος),

who holds his tablets in his left hand, while his master is unfolding a roll of papyrus. The subjects treated on vases of the fourth century B.C. are freer, and less exclusively mythological. The fancy of the artist busied itself with scenes of daily life. A cup from Vulci shows us a veritable idyl. Two young men are conversing with a man of mature age seated upon a rough stool. One of the young men points at a swallow, and exclaims, "Look! a swallow!" (ἰδοὺ χελιδών). "Yes, by Heracles!" (Νὴ τὸν Ἡρακλέα) replies the older man. "It twitters!" (αὐτεῖ) replies the younger lad. "Spring has come" (Ἔαρ ἤδη) is the conclusion drawn by the bearded man. It would be impossible to treat a scene like this with a simpler grace. It might be entitled "The Return of Spring."*

(3) *Vases of the Attic style.*—It is easy to see that the more we advance into the fourth century B.C., the more do we find the pure Attic style tending to prevail in the pottery of Greece proper. At the same period the differences between purely Greek productions and those of Italy grow more distinct. We must, therefore, make a special series of vases of the Attic style, the characteristics of which are strongly defined. This delicate pottery is generally in the form of vases of small size, like the pyxis, œnochoë, aryballus, etc.; they may be easily recognised by the beautiful black of their coating, by garlands of myrtle with pointed leaves often serving as a decoration, but above all by the extreme

---

* *Mon. inediti dell' Inst.*, Vol. II., pl. xxiv. [The inscriptions are IΔOXELIΔON NETONHEPAKLEA IƷTTΑH (reversed) EAPEΔE.]

U

delicacy of their painting. Athenian artists alone

✝

K I S S O

✝ P Y S I S

FIG. 115.—FRAGMENT OF A DIONYSIAC SCENE.
(Upon a red-figured aryballus.)

seem to have possessed the lightness of touch and

exquisite delicacy of style which are shown on this class of vases.

European museums possess some beautiful examples of these vases. One of the most remarkable is the aryballus of the Naples Museum,* which represents the battle of the Amazons and Athenians. Though found at Cumæ, the vase has all the characteristics of the Attic style, and has nothing in common with the Italo-Greek pottery of the same epoch. Qualities purely Attic appear also on the painting upon an aryballus, found at Æxone, showing Dionysus and his train. The gracefulness of attitudes, the facial expressions, and the exquisite lightness of the work, render it a masterpiece. Though Athenian vase-painters continued to receive inspirations from mythology, they also frequently employed motives suggested by every-day life. The collection in the Varvakeion is rich in Attic vases in which scenes of indoor life are figured:† ladies at their toilette, or visiting; the women's apartments, where women are spinning and talking ; scenes of betrothal and courtship, in which winged Loves interpret the sentiments of the

FIG. 116.—ON A SMALL RED-FIGURED LECYTHUS.
(Museum at Athens.)

* *Raccolta Cumana*, No. 239 of the catalogue by Heydemann.
† *Catalogue du Musée d'Athènes*, Nos. 406—499.

U 2

persons represented. The series is large and varied. Artists by no means disdained to adorn with great care the smallest vases, which were used as children's toys, and were decorated with compositions suggested by the sports of childhood. Nothing is fuller of life and spirit than these tiny subjects treated in an animated way, in which children are seen dragging their carts, riding on the back of a dog, or pushing wheels before them with long sticks. It is family life, taken from nature and treated with perfect art.

(4) *Vases of common workmanship.*—These vases are found in great numbers in our museums, and are more or less carelessly executed. They possess no characteristics clearly enough defined to allow of their being classified with one of the preceding series. In general they are of great size, like the hydria, calpis, celebe, crater, or amphora, and to these differences of classes frequently correspond differences of subject. One of the most frequent motives found upon amphoræ is that of betrothal or of wedding processions, at which are present the *nympheutria* and women bearing presents—vases or jewels enclosed in caskets. Pelicæ and hydriæ frequently offer subjects from ordinary life : pictures of interiors, representing women seated and at work, surrounded with tame birds ; sometimes also toilette scenes. Crateræ and canthari are decorated with Dionysiac subjects : Dionysus, crowned with parsley, bearing the thyrsus, and surrounded by Bacchantes and satyrs, who carry canthari and prochooi. This class of subjects is especially common on vases from Bœotia. These vases

may be recognised by their execution, which is sometimes negligent, and by the heavy style of the palm-
leaf ornament which accompanies the paintings. There
are many examples in the Museum of Athens.

(5) *Vases with gilded ornaments, and with reliefs
heightened with gold.**—During the most brilliant
period of ceramic art among the Greeks, the decoration
of vases attained great richness, with the aid of painting in polychrome and of gilding. At the close of the
fourth century B.C. it was a general custom to gild
certain parts of the costume, such as fillets, earrings,
beads in necklaces, berries in laurel garlands or in
myrtle wreaths, which are found on vases as subordinate ornaments. The gilding was done by means of
leaves of beaten gold laid upon low reliefs or upon
small bosses of clay paste. Upon the small vases of
Attic style, such as aryballi and œnochoœ, the gilding
is frequently applied with careful discrimination, but
upon large vases it is profusely lavished. Vivid colours
heighten the effect of the painting, and tints of red,
green, white, and violet, applied to draperies, united
their brilliancy with that of the gold.

One of the most beautiful vases among those in
which gold and varied colours are found together,
is a pelice found at Cameirus, now in the British
Museum. The painting represents the abduction of
Thetis by Peleus. The peplus that the goddess lets
fall is sea-green, with a white border ; her ornaments,

---

* See J. de Witte, *Revue archéologique*, Vol. VII. ; Otto Jahn,
Ueber bemalte Vasen mit Goldschmuck, 1865 ; M. Collignon, *Trois
Vases peints de la Grèce propre à Ornements dorés*, in *Rev. arch.*, 1875.

and those of the Nereids in her train, are richly gilded.
Among the gilded vases of the Museum at Athens we
will mention a pretty aryballus, on which Athena is
represented crowning Pelops after his victory, in
front of the statue of Athena Cydonia, unto whom
the hero had sacrificed before beginning the contest.
This vase, which was found in Attica, is of unusual
purity of style, and is a good example of painted
vases of small size, where gold is applied with the
moderation demanded by the limited proportions of
the subjects figured.

It had been believed for some time, and with
probability, that vases richly decorated with gold were
of exclusively Athenian manufacture; their delicacy
of style seemed to suggest that origin. But as the
number of common vases discovered has increased,
the conviction has grown that these must have been
manufactured in very different places; outside of
Attica, Bœotia, Megaris, and Corinth have furnished
remarkable examples. This luxury in ornamentation
was not peculiar to Athens; it was found, without
doubt, in the principal establishments for the production of vases in the Hellenic world, and the artists
of Corinth, in this respect, could rival those of Athens.

There is a whole class of vases where paintings are replaced by figures modelled in relief,
and applied, as a sort of frieze, to the body of the vase.
Though vases of this kind are rare, they take rank
among the wonders of Greek ceramic art. The
most beautiful of them show the combined employment of all the resources of decorative art—relief,

polychromy, and gilding. We must give a place among the first to the famous Cumæ vase, once in the Campana collection, but now in the Hermitage Museum in St. Petersburg. The groundwork of the vase, covered with a brilliant black coating, is fluted ; the principal group of figures in relief is made up of Triptolemus and the Eleusinian goddesses. In the second frieze are represented lions, dogs, panthers, griffins—likewise gilded. The marvellous work in the modelling, the richness of the gilding, the brilliancy of colour lavished upon garments, make this vase one of the most valuable monuments of ceramic, as well as of plastic, art. Similar qualities distinguish an aryballus found at Kertch in the Crimea. The subject represented is a hunting scene, in which young Darius and his companions, Abrocomas, Seisames, Euryalus, and Eurus take part. The signature of an Athenian artist named Xenophantus is affixed to this vase. The date is about the one hundredth Olympiad (380 B.C.).*

Sometimes relief is combined with painting. This is the case in another vase from Kertch, representing the dispute between Athena and Poseidon. These two divinities, alone modelled in relief, occupy the central part of the composition. In this group may be recognised an imitation of the statues that formed the central subject of the western pediment of the Parthenon.

---

* ΞΕΝΟΦΑΝΤΟΣ ΕΠΟΙΗΣΕΝ ΑΘΗΝ [αιος]. *Antiquités du Bosphore Cimmérien*, 1854. Plates XLV., XLVI. [*Cf.* on many of these vases from Kertch, C. T. Newton, *Essays*, 1880, p. 373 *ff.*]

(6) *Vases with reliefs.*—In this class of vases the reliefs were modélled separately, and applied afterwards.* But we can easily understand that the inventive genius of the Greeks must have imagined other systems of decoration into which reliefs should enter. Such a system is found in vases adorned with bas-reliefs that were produced· by means of a stamp, or of hollow moulds, impressed upon the clay while yet undried. This process is very ancient ; we find it used on vases of a very remote period. Decorated according to this system are cups from Megara, vases hemispherical in form, with black varnish, upon which are represented scenes from the myths of Dionysus. We have already described receptacles in the form of statuettes, or figurines, where the neck and orifice are all that remind one of the principle of the vase.† In this direction there were no bounds to the fantasy of the artist. It would be difficult to specify all the combinations created by whim, or by a taste for novelty.

(7) *White Lecythi from Athens.*‡—It is here proper to bestow some attention upon an extremely interesting class of vases, not found outside of Attica.

---

* This process in the technique may be clearly seen on a vase in the form of a flat gourd in the Museum at Athens ; the bas-relief, representing the battle of the Amazons, has become loosened in places.

† See G. Treu, *Griechische Thongefässe in Statuetten- und Büstenform*, 1875.

‡ See O. Benndorf, *Griechische und Sicilische Vasenbilder*, 1869—1883, and the study especially devoted to this class of vases by E. Pottier, *Les Lécythes blancs attiques à Representations funéraires*, 1883.

These vases are of uniform type, and have very clearly defined characteristics. They are all in the form of elongated lecythi; the body is adorned with a white coating, very bold in colour, easily erased by the finger-nail. Upon this coating are painted figures usually relieved with bright colour. The drawing is made in reddish-brown. A very brilliant black varnish covers the neck and the foot.

Lecythi of this class are often found in Attic tombs, but never elsewhere.* This is explained by the use to which these lecythi were put. They played an important part in Athenian rites relating to the cult of the dead. We know that, together with beliefs commonly held throughout Greece, each district had its special customs, its own funeral ceremonies, and other practices with regard to the dead. Those of Attica were very peculiar, and it can easily be believed that a class of industry belonging

FIG. 117.
WHITE LECYTHUS
FROM ATHENS.

especially to these observances might never have passed the limits of the country. It was only at

---

* These vases, though rare in most European museums, are very common in Athenian collections.

Athens that these vases were made, which as far back as Aristophanes were spoken of as associated with death. The comic poet speaks of one "who paints lecythi for the dead."* Monuments also show that they figured in the ceremony of the laying-out of the dead (πρόθεσις).

The reference in Aristophanes proves, moreover, that white lecythi were manufactured as early as 392 B.C. It is probable that this manufacture continued throughout the fourth century B.C., and did not cease until the end of the third century B.C. There are no specimens of white lecythi of archaic style. The specimens most carefully executed plainly indicate their date as that of the highest perfection of Attic style, and vases of this class are not found in tombs of the Roman epoch. These facts afford sufficient reasons for limiting the manufacture of white lecythi to two or three centuries.

Subjects figured on these vases have in common the characteristic of being suggested by customs connected with burial. With the marble stelæ, they are the most important and trustworthy materials for the study of the history of the cult of the dead in Attica. The traditional form of these customs, and the respect of the Athenians for rites that had to do with their deepest feelings, did not permit the ceramic artist any great variety of subjects. The paintings upon lecythi confine themselves mostly to the following subjects :—

Offerings at the stele of the dead ; the lamentation.

---

* Aristophanes, *Eccles.*, 995.

—The scene of the offering is ordinarily conceived in a very simple manner.* From each side of the stele adorned with fillets approach persons holding in their hands the objects with which they would do homage to the dead, viz., tæniæ or fillets with which to decorate the stele, a canistrum or flat basket

FIG. 118.—THE OFFERING AT THE STELE.
(White Lecythus from Athens.)

with fruits and cakes, sometimes a fowl, as on a lecythus in the Varvakeion in Athens.* (Fig. 118.) Not infrequently one of the persons represented is an ephebus clad for war, or for a journey, who seems to have returned to fulfil the due funeral rites at the tomb of a parent. This scene may also be more complex. Thus, upon a beautiful lecythus† a young lad holding a lyre advances towards

* No. 661 in the *Catalogue*.   † No. 650 in the *Catalogue*.

the stele, led by an older man wearing a chlamys of sombre hue. The presence of the soul of the dead is indicated by a curious detail. A small winged figure (the ψυχή or the εἴδωλον of the dead) hovers near the monument, as if moved by the sound of the lyre. We were already aware that Attic genius did not shrink from treating the conception of death upon the stelæ, but did it with great moderation. Lecythi confirm the testimony of stelæ on this point. On several vases the dead person herself seems to receive the offerings. She is represented as a woman on whose face is an expression of sadness. Resting upon her elbow, which lies on the low back of her chair, she is seated at the base of her own monument, and seems to accept the gifts which living friends are bringing unto her. This is the subject represented upon a polychrome lecythus of exquisite style.* The person seated at the foot of the stele has her head inclined, and the face has a peculiar charm in its features. A woman bringing offerings turns to look at a little girl who follows.

The scene of the *lamentation* is only a variation of the preceding. It represents persons grouped about the stele in attitudes of sorrow, and with hands raised to the head making the significant gesture of mourning. Certain details deserve to be noted. One vase shows us a woman kneeling upon the steps of the stele, with body bent forward as though she were speaking with the dead, and were reproaching him

* No. 637 in the *Catalogue*.

for having left her.* Even at the present day, in some districts of Greece, the custom of chanting *myrologues* or lamentations for the dead is still preserved.

The toilette.—At the foot of a mortuary monument sits the dead woman, receiving homage. Her female attendants bring to her her articles of attire. The character of these representations, generally very thoughtful, shows with what delicacy Attic genius aimed to alleviate the sadness of the idea of death. A white lecythus, a masterpiece of ceramic art, presents this scene of the funereal toilette with rare beauty. The dead, richly robed, throwing back her veil with a graceful gesture, seems to return once more to terrestrial life. Her favourite bird is perched upon the back of her hand; her women bring her a fan, with vases of perfumes. Did not the stele remind us of the funereal intent of the painting, we should believe ourselves looking upon a scene in ordinary life.

The laying-out of the dead.—Here the idea of death appears in all its reality.† The dead man, crowned with flowers, lies upon his funeral bed; fillets and garlands adorn the ground of the vase, suggesting the decoration of the hall of mourning. Near the bed a large lecythus makes known to us the

---

* *Cf.* Benndorf, *op. cit.*, plate XXIV. 1—3, "Women tearing their hair before a monument." The painting is careless, but the expression is very truthful.

† One of the best specimens is a polychrome lecythus of the Museum of Vienna. O. Benndorf, *op. cit.*, plate XXXIII.

part played by this vase in the ceremonies of the
πρόθεσις. The relatives of the dead are yielding to
demonstrations of grief. Sometimes a small winged
figure flies near the bed, image of the living breath or
soul just breathed from the body. All the details of
the scene are drawn with precision. The funereal rites
of Athenians would leave nothing to accident. The
provisions of the law on these points are well known ;
such as the decree of Iulis at Ceos, which regulated
minutely all funeral ceremonies, and the decree of
Solon, declaring obligatory the rites of πρόθεσις.
Ceramic painters in these designs were only reproducing scenes familiar to all Athenians.

Burial in the tomb.—This subject is rarely found
upon white lecythi of Athens; and yet one of the vases
representing it is a perfect masterpiece. Two winged
genii sustain with great care the body of a young
woman, which they are about to deposit in the tomb
excavated at the foot of the monument. A young
man, standing near the stele, looks upon the scene
with sadness. The image of the reality gives place
to an interpretation of the idea of death, treated
with the purest taste. It is an ideal scene, to which
the marvellous excellence in style, the attitudes of
the figures, and the graceful lines of the body of the
young maiden, who seems asleep, have lent an
enduring charm. (Fig. 119.)

Charon and his bark.—We find here once more
the beliefs of the current mythology. The Museum
of Athens possesses several fine instances of this
scene, where Charon, leaning upon his oar, wearing a

sailor's cap, is making ready to receive into his bark persons standing at the water's edge.

FIG. 119.—BURIAL.
(White Lecythus from Athens.)

Other subjects, less frequently found, are the scenes of the *Farewell*, so often figured upon marble ; of the mounted soldier fighting with an enemy on foot ; of figures of divinities, as of Demeter and Triptolemus ; also *Epitaphia*, or scenes commemorative of funeral ceremonies.

It will be observed that all these scenes depicted upon vases are intimately related to the Greek views of death. They afford invaluable aid in the study of marble stelæ, for they are inspired by the same beliefs and sentiments that sculptors had rendered upon funereal bas-reliefs. The paintings on lecythi are most expressive and suggestive. While conventionality may have had a certain share in sculptures that were under the control of the traditions of art, it is less evident in these paintings, which were often executed by indifferent potters, to satisfy the demands of popular beliefs.

The style of these paintings is also of peculiar interest. We have already mentioned the most beautiful specimens, those of more finished execution, that may be compared with stelæ of the best style. But many of these paintings show indifferent workmanship. This is true of the majority of them. They deserve, nevertheless, to be closely studied ; imperfect as they may be, they still strikingly remind us, among other things, of the Athenian type, naïvely rendered, such as the long nose, the strongly marked chin, all the features that archaic masters copied with such care, and that disappeared from sculpture in the classical epoch. At the same time, underneath the carelessness of the work, is distinctly felt the artistic tradition, that impressed itself upon the humblest draughtsman. In the attitudes of the figures, in the arrangement of the draperies, we find the marks of the noble style. No examples could better show how popular was art in Athens, and how it found a place even in the smallest creations of this gifted people.

We cannot follow ceramic art into the period of decline, represented principally by products in Magna Græcia. The vases of Sant' Agata dei Goti, of Ruvo, and of Armento, in Italy, show how far the exaggeration of forms, the careless use of colours and of ornaments, and the taste for the bizarre, rapidly led this species of industry away from the simple and noble traditions that had given to it its high honour.

V

CHAPTER IV.

TERRA-COTTA PLAQUES WITH PAINTINGS.*

WITH the study of painted vases is closely connected that of terra-cotta plaques decorated by the same process that was used in ceramic paintings. This class of monuments has been studied but a short time, and the rarity of specimens known gives it great importance. These painted plaques are oblong in shape, and are covered with figured subjects. The same order in styles that we have already noted is found here also—paintings with black figures, and with red figures.

Black-figured plaques are the most numerous. We mention as a type of this class a specimen from the Louvre, representing the laying-out of and the mourning for the dead.† Figures of persons in the archaic style are designated with scrupulous care by accompanying inscriptions. At the head of the bed are the grandmother (θεθε), the mother (μετερ), and a sister (αδελφε); further along two women (θετις προς πατρος, "maternal aunt," and τεθις, "paternal aunt") make gestures expressive of sorrow, while

---

* O. Benndorf, *Griechische und Sicilische Vasenbilder*, 1869–1883; Dumont, *Peintures céramiques*, p. 29 *ff.*

† Benndorf, *op. c.*, plate I.

the father and the brothers standing at the foot of the bed abandon themselves to lamentation. Exclamations of grief are expressed in inscriptions written on the field (οἴμοι, " alas!"). The same subject is found on two other plaques of similar style.

FIG. 119*b.*—MINERS AT WORK.
(Painted Terra-cotta Plaque. Berlin Museum.)

Other examples present mythological scenes, Heracles and Iolaus, Athena in a chariot, etc. Finally, some painted tablets, lately discovered at Corinth,* represent Poseidon, and scenes taken from ordinary life,

---

* O. Rayet, *Gaz. arch.*, 1880. The Berlin Museum possesses numerous specimens of these votive tablets. [Cf. Furtwängler, *Catal.*, Nos. 347–955.]

V 2

such as pugilistic contests, potters occupied in firing
their vases, miners with pickaxes working at the
base of a mountain.

Among red-figured plaques, which are very rare,
we would make especial mention of fragments that
show processions of the gods, Hermes, Apollo with
a lyre, and Athena armed with a spear.

The object of these monuments is not a matter of
doubt.   They are unquestionably votive offerings
(*ex voto*) dedicated either in tombs or in temples.
Certain vase-paintings explain their use.   Thus an
amphora in Munich exhibits an ephebus or a victorious
athlete, holding sprigs of myrtle and a painted
plaque like those we have described, representing a
runner. The young victor is about to consecrate in a
shrine the *ex voto* offering that indicates his triumph.
A still more decisive witness is furnished in the inscriptions sometimes to be read on these monuments.
Among some black-figured plaques found near Corinth
several bear a dedication to Poseidon.   Thus one
lately acquired by the Berlin Museum, signed Timonidas, on one side shows Poseidon near the potter's
oven, on the other a hunter with his dog.   The
plaque is inscribed [Ὁ δεῖνα] ἀνέθεκε τôι Ποτειδᾶνι
" [Such an one] dedicated me to Poseidon."   Some of
these plaques have been found uninjured.   They seem
to have been thrown outside of the temple enclosure
as rubbish, and were doubtless removed in order to
make place for more important votive tablets.

Book Sixth.

NUMISMATICS AND GLYPTICS.

CHAPTER I.

NUMISMATICS.

ECKHEL: *Doctrina Nummorum Veterum*, 1792-1798.
MIONNET: *Médailles grecques et romaines*, 1806-1838.
BEULÉ: *Monnaies d'Athènes*, 1858.
FR. LENORMANT: *La Monnaie dans l'Antiquité*, 1879 (unfinished).
IMHOOF-BLUMER: *Monnaies grecques*, 1883.
PERCY GARDNER: *Types of Greek Coins*, 1883.
BARCLAY V. HEAD: *A Guide to the Principal Gold and Silver Coins of the Ancients from* B.C. 700—A.D. 1. Second Ed., 1881.
BARCLAY V. HEAD: *The Chronological Sequence of the Coins of Bœotia*, 1881.
PERCY GARDNER: *Samos and Samian Coins*, 1882.

THE history of ancient money, the principles that ruled in coinage, and the study of types figured upon coins and of accompanying inscriptions, constitute a special science—that of Numismatics. We cannot here give even an outline of it, but must confine ourselves to indicating the relations that connect the science of coins with the archæology of art. This science is an indispensable aid in political and social history, in geography, and in other studies, and is equally useful in the history of art. Since, in 1842,[*] Raoul Rochette showed the advantage that might be

---

[*] *Mémoires de l'Académie des Inscript. et Belles-Lettres.*

derived from a knowledge of coins in the restoration of statues described by ancient writers, or in the recognition of copies in the marbles of our museums, the application of numismatics has come into very general use. Applied either to statues or to monuments of architecture,* this method has been most fruitful in results, in the comparison of types on coins with descriptions of ancient authors and with figured monuments.

Considered in themselves, coins are of profound interest in the history of art. They form a chronological series, rich and varied, in which may be followed with great accuracy the progress and decline of style. From this point of view, coins and medals may be classified according to great periods, corresponding to the principal epochs of Greek art. It will be useful to indicate, in each one of these divisions, the most remarkable types of coins found in it.

I.—The most ancient Greek silver coins seem to date from the eighth Olympiad (748 B.C.). Pheidon, king of Argos, is credited by ancient writers with the invention of coinage. The mints of that time struck off pieces of irregular shape, having the appearance of small ingots, the convex surface of which alone bore an emblem. The reverse is marked by a square and deep indentation (*quadratum incusum*), caused by the square head that served to hold the piece when receiving the impression.

---

* *Cf.* Donaldson, *Architectura Numismatica*, 1859 ; Lenormant, *La Numismatique et l'Architecture* in *Revue génér. de l'Architecture*, 1877.

The emblems on the coins are somewhat complex. At Ægina, a tortoise ; at Ephesus, a bee ; the winged Pegasus and the letter *koppa*, at Corinth ; a Gorgon's head at Athens ; a shield on Bœotian coins. No legend is found with the emblem on these primitive coins.

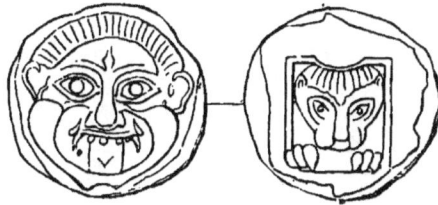

FIG. 120.—SILVER COIN OF ATHENS.
(Most ancient style.)

II.—Between 580 B.C. and 460 B.C., the progress of art made itself felt in the coins produced. The very simple emblems of the earlier age are succeeded by heads of divinities or of mythological persons, while the incuse square on the reverse is gradually filled with letters and with designs. Of this sort are the coins of Ægina, of the transition style, where the obverse has the tortoise with the letters *AI* ; on the reverse is the incuse square with a dolphin and the letters *IN*: Aἰ[γ]ιν[ατᾶν].

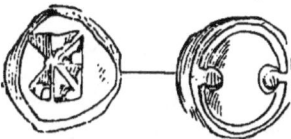

FIG. 121.
BŒOTIAN SILVER COIN.
(Most ancient style.)

The same progress may be followed on Attic coins ; those of Solon and of Peisistratus have on the reverse the incuse square crossed by diagonals, and on the obverse emblems like the Gorgon's head, part of a horse, a wheel. After the expulsion of the Peisistratidæ, the head of Athena, in archaic style, appears, and is reproduced for a long time subsequently : this type is treated in the manner peculiar to archaic art. The eye is drawn as if in front, and the vigorous lines of the profile have great

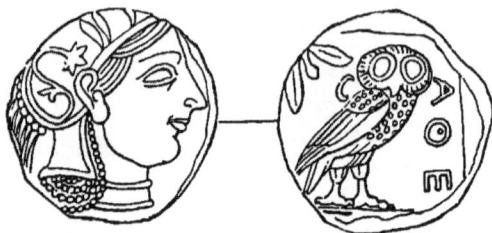

FIG. 122.—ATHENIAN TETRADRACHM.
(Archaistic style.)

severity. Upon the reverse is the owl, the seal of the state, with a crescent, a branch of olive, and the legend $AΘE$: ᾿Aθη[ναί- ων].

To this period belong the incuse coins of Magna Græcia, the obverse of which bore a subject in relief that appeared incuse on the reverse. At this time were struck coins of Rhegium and Messina, with mules yoked to a car, and a hare for types ; those of Gela, with the river god having a human head, but the body of a bull ; also ancient coins of Syracuse bear- ing the head of

FIG. 123.—COIN OF SYRACUSE.
(Archaic style.)

Arethusa surrounded by dolphins, and on the reverse a *biga*, or two-horse chariot, above which hovers a figure of Victory.

The progress is clearly marked on the coins of

Thasos : the earliest exhibit a satyr carrying off a nymph ; the treatment is of a rude style that recalls the crudeness of the conceptions of primitive Greek art. Upon the coins of the following age the scene is rendered with much more refined sentiment, and the short thick figures give place to more elegant and more correct designs. The beautiful coins of Macedon and of Chalcidice resemble those of the era of perfection. As we advance in time the inscriptions upon coins become explicit and definite. The earlier legends confined themselves to a few letters, sometimes to only one, indicating the name of the city (ΘA for $\Theta a[\sigma i\omega\nu]$, coins of Thasos ; $E\Phi$ for $'E\phi[\epsilon\sigma i\omega\nu]$, of Ephesus ; A for $'A[\rho\gamma\epsilon i\omega\nu]$, of Argos, etc.). Subsequently the practice prevailed of writing in full all the letters of the name of the town, although there are some exceptions to the rule.

III.—This epoch, comprising the time between 460 B.C. and 336 B.C., shows the acme of the art of coinage. During the former half of this period, the style of coins, though very pure, retains some traces of archaism. As already remarked, the coins of Athens retained until the best period the ancient type of Athena. The coins of Sicyon with the chimæra, of Argos with the wolf, and of Agrigentum with two eagles standing upon a hare, well represent the first movement in the period.

At the end of the fourth century B.C. the coins issued by Greek cities are characterised by the highest style of art. Few of them are more finished than the beautiful coins of Pheneus in Arcadia, with the head

of Demeter or of Core on the obverse, and on the reverse Hermes carrying the youthful Arcas. The coins
of Stymphalus, with
the head of the Arca-
dian Artemis, and
Heracles contending
with the Stymphalian
birds, are master-
pieces. The magnifi-
cent development of
Peloponnesian sculp-

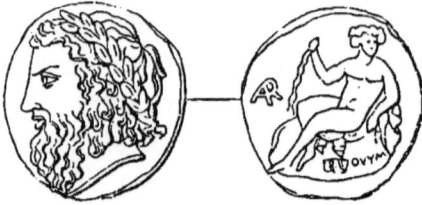

FIG. 124.—SILVER COIN OF THE
ARCADIAN CONFEDERACY.

ture exercised a powerful influence upon the art of
coinage in Arcadia at the time when the Arcadian
Confederacy caused the coins to be struck that
asserted its rule
over Olympia. This
league adopted for
symbols the head
of Olympian Zeus,
and the figure of
the god Pan, seated
upon the summit
of Mount Lycæum,

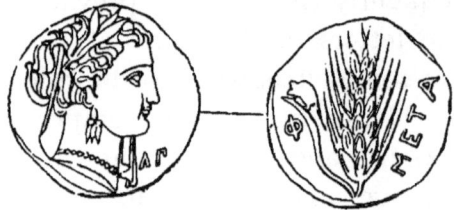

FIG. 125.—SILVER COIN FROM
METAPONTUM.

the Arcadian Olympus, as the legend $OAYM$
(Ὀλυμ[πος]) indicates.

In Sicily and in Magna Græcia coinage attained
a degree of perfection that has never been surpassed.
Engravers signed their coins as sculptors their
statues.*

* A collection of the signatures of engravers of coins has been
made by Von Sallet, *Die Künstlerinschriften auf griechischen Münzen,*

To this period belong the coins of Catana and of
Sicilian Naxos, signed Procles; those of Metapontum,
cut by Aristoxenus; and especially the admirable
pentacontalitræ of Syracuse, signed by Cimon and
Euænetus, where the head of Arethusa is treated with
rare nobility of feeling.

Artistic progress is apparent not only in the great
excellence of the types of divinities, that often repro-

FIG. 126.—ATHENIAN TETRADRACHM.
(Later style.)

duce the works of great masters, such as the Hera of
Polycleitus and the Olympian Zeus of Pheidias—the
engravers of coins and medals also treat with rare
genius, in the limited field of the reverse, genuine
plastic scenes. Victories won at the Olympic games,
games celebrated in honour of the gods, mythological
legends relating to a city, are the ordinary subjects of
these compositions. Thus the reverse of Syracusan

1871. Few signatures are found outside of Sicily and Magna Græcia,
and in these countries they are limited to a short period—from 490 B.C.
to 350 B.C.

coins exhibits a four-horse chariot, driven by a charioteer who receives a crown from a Victory, and in the field are the weapons given as prizes to the victors. On the coins of Selinus, Apollo and Artemis advance, Apollo with stretched bow, as divinities of the plague ; on the reverse the river-god Selinus ($ΣΕΛΙΝΟΣ$) makes offerings to Asclepius to stay the scourge. Here is an evident allusion to an

FIG. 127.—SILVER COIN OF SYRACUSE.

incident in the history of the city. At the advice of Empedocles, the people of Selinus changed the course of the rivers Selinus and Hypsas in order to drain the marshes surrounding the city.*

Another characteristic of this period is that engravers represented on coins the head in front view or in three-quarter view, executed in high relief.† About the eightieth Olympiad (460 B.C.) Cimon of Cleonæ was the first to break with the timid traditions of earlier painters. He represented

* Diog. Laert. viii. 2. 11, 70.

† Cf. De Witte, Médailles d'Amphipolis, in Revue numismatique, Vol. IX., 1864.

heads in front view—a thing that Polygnotus and Micon had not dared to attempt.   The engravers of coins applied this innovation to monetary art about the time of Alexander, tyrant of Pheræ (369 B.C.), who caused a beautiful tetradrachm to be struck bearing the head of Artemis in front view.   Many Greek cities followed his example.   The head in front view appears on the coinage of Amphipolis,

FIG. 128.—SILVER COIN OF ANTIOCHUS IV.

Thebes, Syracuse and Catana, Metapontum, Croton, Rhodes, Clazomenæ in Ionia, and of other cities. But this practice lasted only a short time.   It was seen that coins in high relief were much exposed to wear, and soon became defaced.   A return was made to faces in profile about the time of Alexander the Great.

IV.—In the epoch of the Macedonian dynasties the art of coinage, as well as the other plastic arts, entered upon a decline.   The coins of Alexander and of his earlier successors are still very beautiful. Those that were struck by Philip Arrhidæus, by

Antigonus, and by Lysimachus, are worthy of the best period of Greek art. In Sicily the coins of Agathocles, of Nicetas, and of Pyrrhus, also testify to fine and pure taste preserved by tradition. But the dynasties of the Seleucidæ and of the Ptolemys struck coins in which the style gradually underwent a change. From the time of Alexander the custom became established of representing a likeness of the Sovereign upon the obverse, while on the reverse were inscriptions setting forth in prolix language the titles and names of the princes. Thus the coinage of Antiochus IV. is inscribed *ΒΑΣΙΛΕΩΣ ΑΝΤΙΟΧΟΥ ΘΕΟΥ ΕΠΙΦΑΝΟΥΣ*. Although these coins are of great value in the history of the Greek dynasties and for the portraits of the rulers,[*] they awaken but slight interest in connection with the history of Greek art. Decadence in taste is to be equally observed in the coinage of the cities in Greece proper.

---

[*] Imhoof-Blumer, *Porträtköpfe auf antiken Münzen hellenischer und hellenisierter Völker*, 1885.

CHAPTER II.

GLYPTICS.

KRAUSE: *Pyrgoteles, oder die Steine der Alten*, 1856.
KING: *Antique Gems and Rings*, 1872.
BILLING: *The Science of Gems, Jewels, etc.*, 1867.
WESTROPP: *A Manual of Precious Stones and Antique Gems*, 1873.

THE study of Glyptics, or of engraved precious stones, is naturally connected with that of coins : its aim is the interpretation of small subjects which often have their analogues on coins. At the same time difficulties are here encountered. The absence of legends or inscriptions makes a chronological classification less certain. We are often reduced to considering simply the style of the engraved stone, except when the subject is of an historical character, and thus indicates a definite epoch. The signatures of engravers, when they can be regarded as authentic, likewise furnish useful indications.

Glyptics is one of the branches of ancient art that most keenly aroused interest at the outset of archæological studies ; collectors have for a long time eagerly sought for carved stones, and since the work of Æneas Vicus of Parma, engraver to Alphonso II., Duke of Ferrara, treatises on ancient glyptics have frequently been published.* This interest gave rise to the

* See Mariette, *Traité des Pierres gravées*, 1750 ; and De Murr, *Bibliothèque dactyliographique*, 1804.

industry of counterfeiting, and the number of spurious gems is considerable. Accuracy in vision, and skill in criticism, are required to distinguish these imitations from authentic works. The European museums possess rich collections of engraved stones ; those of Paris, Florence, Vienna, Naples, and Berlin occupy the first rank.

I.—From the point of view of technique, engraved stones are divided into two grand classes, *intaglios* and *cameos*. The former class includes stones in which the engraved subject is sunk beneath the surface ; from them is obtained in soft wax an impression in relief. For this class, stones of a single colour are chosen, such as the amethyst, hyacinth, agate, cornelian, chalcedony. The polisher gives the stone an oval and convex form, and the engraver works at it with the chasing-punch, taking the greatest pains to polish all the cavities of the figure he traces. This finish in work is regarded as a sign of authenticity.* Imitations of intaglios were not rare, even in antiquity. False gems were made by taking impressions in pastes of coloured glass ; these were then set in rings, necklaces, and other jewels, and were the luxury of the poorer classes. Most commonly intaglios served as seals or signets (σφραγίς), and were worn in finger-rings.

Cameos are distinguished from intaglios by being engraved in relief; they are often of very large size. Engravers frequently cut them in stones where

---

* See the technical details given by Pliny, *N.H.* xxxvii.

many colours are found in several layers, and thus obtained a different coloration for the relief and for the ground.

II.—Ancient Greek intaglios seem to have been engraved in imitation of Egyptian scarabs, and of Babylonian cylinders. They have been found in the tombs at Mycenæ, where they cannot be regarded with certainty as Hellenic in origin ; relations with the East must very early have furnished models to Greek engravers.* It would be superfluous to return to the question as to Oriental sources, and to indicate at length the Græco-Oriental characteristics of many ancient intaglios. These characteristics distinctly appear on stones found at Cyprus, in the " Curium Treasure."† The style of engraved

FIG. 129.—APOLLO WITH THE DOE.

* On the archaic intaglios found on the islands of the Ægean Archipelago ("island gems"), see especially Milchhöfer, *Anfänge der Kunst in Griechenland,* 1883.

† Engraved stone representing Nemesis : *Gazette archéologique,* 1878 ; Myrtilus and the horses of Œnomaus, *ibid.*

W

stones passes through the same phases as that of sculpture. Gem-cutting, represented in the primitive period
by the Samian Theodorus, who made the famous ring
of Polycrates, was developed as a branch of sculpture ;
it attained perfection in Pyrgoteles, who engraved the
seal of Alexander, and did not cease, even in the
Roman Empire, to produce the delicate pieces of
work that adorn
our museums. The
artists whose signa-
tures are found on
the best known
gems belong en-
tirely to the Mace-
donian period and
to the Roman
epoch, if we are to
judge by the shapes
of the letters and

FIG. 130.--GEM OF APOLLONIDES.

by the formulæ used : Satyreius flourished under
Ptolemy II. ; Tryphon under King Polemon, the
protégé of Mark Antony ; Dioscurides under Augustus.
The names are known of a large number of engravers
belonging to this period, but we cannot admit
with certainty all the names that are to be
read on gems. Counterfeiters have often placed upon
antique stones fictitious names to give them greater
value,* or have actually inscribed names that they

---

* See Letronne, *Bull. de Corr arch.*, 1851. On one gem is to
be read the name Πωημου, which is not Greek. In another place the
counterfeiter has incorrectly written a name : Αλλιον for Δαλιων.

have misunderstood. Accordingly the signatures of engravers must be accepted with great reserve ; those that are the least doubtful are the following : — Agathapus, Apollonides, Aspasius, Athenion, Boëthus, Dioscurides, Epitynchanus, Euodus, Eutyches, Phelix, Heracleidas, Herophilus, Hyllus, Cœnus, Mycon, Neisus, Nicandrus, Onesus, Pamphilus, Protarchus, Solon, and Teucrus.

FIG. 131.
BETROTHAL SCENE.
(Engraved Gem.)

Great variety as to subjects treated prevails among the engravers of precious stones. We can easily convince ourselves of this by running over the catalogue

FIG. 132.—PELOPS WATERING HIS HORSES.
(Cameo.)

prepared by Chabouillet of the engraved stones in the French Cabinet des Médailles. Mythological

W 2

subjects are very common. Often artists reproduce
the classic types of Greek divinities. Such are the
beautiful intaglios of Eutyches and of Aspasius, representing the head of Athena. The gem of
Aspasius is of so pure a style that many scholars,
with Eckhel, have regarded it as a copy of the chryselephantine Athena of Pheidias. On the cameo of
Zeus Ægiochus, found in Ephesus, the god has a
calm expression of power that recalls the best traditions of art. It would be easy to multiply examples, and to show that engraved gems are a very
rich source of knowledge in the study of mythology
as illustrated in art.

Engravers did not confine themselves to the representation of heads ; in the limited field of the gems
they attempted complicated scenes, and drew inspiration from the most attractive subjects in mythology.
The cyclus of Eros, and that of Aphrodite in
particular, set in play the artists' imagination, and
furnished them with charming compositions, where
a felicitous conception is allied with most delicate
execution. Into plastic forms of the most exquisite
beauty, engravers often translated the epigrams of
the poets of the Anthology. There seems to be a
natural bond connecting these lighter products of art
and poetry. In an epigram, the poet Crinagoras
addresses Eros in chains as one punished for his
misdeeds :—" Yes, cry and sob, thou traitor, now that
thy hands are tied ; it becomes thee to weep. No
one shall release thee. Do not look to one side.
Thou hast made many eyes glisten with tears,

shooting thy arrows into hearts, while thou distillest
the poison of desires from which one cannot escape,
O Eros! The sorrows of mortals seem to thee
matters for laughter. Very well, now thou sufferest
as thou hast made others suffer. There is good in
justice."* May we not believe that in writing these
verses the author had before his eyes a beautiful in-
taglio, that represented Eros disarmed, his hands tied
behind his back, and in confusion because of his mis-
hap? On another gem Eros is torturing Psyche, the
image of the soul, by burning it with his torch. This
motive is found in an epigram of Meleager :—" If thou
persist in burning the unhappy Psyche, she will flee, O
Eros, for she too, thou wretch, has wings."† Engraved
stones make known to us a whole phase of Greek
thought that was developed principally in the Macedonian epoch, where are observed refinement in taste
pushed to playful affectation, extreme subtlety, and
rare facility in putting into ingenious forms the fine
analysis of sentiments, by which at that time art and
literature alike were largely nourished.

An important class is that of gems adorned with
portraits ; masterpieces of Greek glyptics are found
among these portrait gems. It is sufficient to mention
the magnificent Gonzaga cameo (now in Russia), representing the heads of Ptolemy II. and Arsinoë ; that
of Vienna, also with the heads of a Ptolemy and of
an Arsinoë, hardly falls behind the Russian cameo in
beauty. These gems belong to the period when

---

* *Anthol. Plan.* iv. 199.　　　　† *Anthol. Pal.* v. 57.

engraved stones were of considerable size ; they were no longer simple seals, and were used in the decoration of costly vases and cups. The ornamentation of gems became more and more elaborate. At the Roman epoch the gems were covered with actual historical scenes, conceived as pictures ; such is the celebrated cameo of Sainte-Chapelle, in the Bibliothèque Nationale in Paris, representing the apotheosis of Augustus. But these complicated compositions, though the work of Greek artists, belong by the spirit that inspired them to the history of Roman art.

𝕭𝖔𝖔𝖐 𝕾𝖊𝖇𝖊𝖓𝖙𝖍.

BRONZES AND JEWELS.

CHAPTER I.

BRONZES.

FRIEDERICHS : *Kleinere Kunst und Industrie in Alterthum. Die Bronzen*, 1871.
FURTWÄNGLER : *Bronzefunde aus Olympia*, 1879.
Compare an interesting study by GUILLAUME, *La Sculpture en Bronze*, 1868 ; also Daremberg and Saglio's *Dictionnaire des Antiquités grecques et romaines*, 1873 *ff.* (in process of publication), *s.v. Cælatura* (by SAGLIO).

WE have already spoken of the place of bronze in Greek plastic art. Bronze was always in favour among the Greeks ; we meet with it in early times in the schools of Oriental Greece, especially at Samos, as well as elsewhere throughout the most flourishing periods of art. The great masters in Greek art excelled in the art of using bronze ; Pheidias, Polycleitus, and Lysippus were toreuticians as well as Calamis, Ageladas and Canachus, and wrought as skilfully with metal as in marble: It even seems that certain schools, like those of Argos and of Sicyon, were traditionally devoted to the art of working in bronze. In the latter years of Hellenism, under the successors of Alexander, this art was exercised by artists with great skill, and the bronzes of Pompeii and of Herculaneum, in which one recognises the hand of

Greek workmen, prove that the traditions of the schools were preserved unimpaired.

This predilection of the Greeks for the art of working in bronze is explained by the qualities demanded by metal working, which in every point coincide with the peculiar aptitude of Greek genius. Bronze requires definiteness of outline and perfect elegance in silhouette effect, while at the same time it allows figures more independence and variety ; accessory parts, such as supports, are not needed, and by means of this omission contours have all their value.

We shall not here return to a study of the schools of bronze-founders, nor to a consideration of the primitive processes employed, at the beginnings of Greek art, in the Oriental schools. Nor shall we attempt to give a review of the knowledge to be obtained from a study of the great bronzes, masterpieces of the toreutic art, which have their definite place in the history of sculpture. We shall only consider bronzes of small dimensions, statuettes, plaques in repoussé, furniture, weapons, etc.—everything, in short, that can be placed under the head of " minor art and industry." At the same time, however humble might have been the purpose of these objects, they not only are important as valuable materials for the history of ancient life, but are also often of great interest in the history of art. Nowhere has the alliance between art and industry been closer than among the Greeks ; this people could not conceive of them as two distinct things.

Articles in bronze are of infinite variety, both in

form and in use. We shall limit ourselves to
noticing three important groups, viz., statuettes, objects
for ornament, and articles for the toilette, among
which mirrors occupy an important place.

§ I. BRONZE STATUETTES.

Greek bronze figurines of small dimensions seem
to have had various uses, indicated partly by the
nature of the subject, partly by their technique ; they
served as images for worship, as votive offerings, as
amulets, or were designed simply for the ornamentation of dwellings.

The votive character of a large number of small
bronzes is not doubtful, and is attested as much by
inscriptions accompanying these articles as by the
circumstances in which they were found. An archaic
statuette of Apollo, discovered at Naxos, has a dedication to the god himself: "Deinagores has conse-
crated me to Apollo, the Far-darter" ($\Delta\epsilon\iota\nu\alpha\gamma\acuteo\rho\eta\varsigma$
$\mu'\alpha\nu\acute\epsilon\theta\eta\kappa\epsilon\nu$ $\acute\epsilon\kappa\eta\beta\acuteo\lambda\omega$ $'\Lambda\pi\acuteo\lambda\lambda\omega\nu\iota$). Certain inscriptions
prove that at times the statue of one god would be
dedicated to another god ; we read an inscription to
this effect upon the bronze Apollo Piombino in the
Louvre, which was consecrated to Athena.* One of
the richest collections of Greek bronzes, that of
Carapanos, is formed of objects excavated at Dodona,
on the site of the temple of Dodonæan Zeus ; it

---

* "Charidemus [has consecrated] to Athena [this statue], the product of a tithe." See De Longpérier, *Notice des Bronzes du Louvre*,
1868, No. 96.

makes us clearly understand the use of these small figurines, which include an infinite variety of subjects, and can have had but one meaning, that of offerings consecrated to the god by devotees who had visited his sanctuary, or had consulted his oracle.*

Certain small bronzes, found in tombs, seem to have been worn as amulets; this is shown by the ring that accompanies them, by which they were hung round the neck.

The decorative intent of bronze statuettes is attested by discoveries in Pompeii. In the villa of Herculaneum one of the rooms was adorned with busts of celebrated philosophers, as Zeno, Epicurus, etc.

We group these figurines according to the subjects they represent, and follow a chronological order in each group; this is the method followed by De Longpérier in his learned *Notice des Bronzes du Louvre*, 1868. We shall limit ourselves to considering them from the point of view of those differences in style which coincide with the progress of art. Avoiding a complicated classification, we may divide small bronzes into two groups :—(1) bronzes of the ancient or antique Greek style ; (2) bronzes of the period of fullest artistic development.

I.—*Bronzes of the ancient style.*†—Among the archaic monuments in bronze of the sixth century

* Several years ago there were discovered at Cape Tænarum a number of bronze horses and other animals, dedicated to Poseidon.

† Greek bronzes of the ancient style are often confounded with Etruscan bronzes; the latter have, however, very marked features, which enable us to recognise them in spite of the resemblance between the two classes.

B.C., the most remarkable are those that were found at Dodona.* The first is a satyr with the feet of a horse, modelled in vigorous style ; he is dancing,

FIG. 133.—PERSON IN ROYAL ATTIRE.
(Bronze from Dodona.)

and his right hand rests upon his hip. In the expression of merry sensuality in his face, in the carefully finished execution of his beard and hair, we recognise the naïve conscientiousness of Greek art at its beginning. De Witte does not hesitate to date this

* Carapanos, *Dodona et ses Ruines*, Paris, 1878.

bronze from the sixth century B.C., or even from the seventh century B.C. To the same epoch belongs the " Flute-player" (*auletria*), likewise found at Dodona, closely wrapped in a tunic of very fine material which reaches to her heels; her mouth is covered with the leather that kept the double-flute in place; she plays upon her instrument in a very realistic attitude. The art of the sixth century B.C. also produced the interesting statue of a seated man, doubtless a royal personage : his hair is braided, and his head covered by the conical cap worn by Thracians ; he is entirely covered by a full mantle, under which one of his hands appears raised as high as his breast. (Fig. 133.)

Although the bronzes of Dodona have greatly enriched the list of archaic statuettes, this list comprised, even before these discoveries, many interesting examples. The Apollo dedicated by Polycrates,* of the old Pourtalés collection, often reproduced by the engraver, shows much of the stiffness of the ancient style.

Bronzes found in the substructures of the earlier Parthenon, which was burned by the Persians, are of equally authentic character, and are thus of infinite value. Such is the statuette of Athena of the old Oppermann collection, which shows the goddess ready for combat, and of which the hieratic attitude and awkward movement remind us of ancient *palladia.*

The transition from primitive art to the freer style of the fifth century B.C. is as clear in the minor

---

* Inscription on the base, in archaic letters : Πολυκράτης ἀνέθηκε.

bronzes as in sculpture on a larger scale. A small bronze Apollo, found at Athens,* appears to show all the characteristics of Attic style about the time of the Persian wars. But the most remarkable specimen of the art of bronze in the earlier half of the fifth century B.C. is without doubt the beautiful statuette of Heracles which once belonged to the Oppermann collection, but now enriches the French Cabinet des Médailles. " The god walks with a rapid movement, his left leg forward and the right in the rear, his calves tense and his knees stiffened ; with his right arm he raises his club, and is about to smite the enemy before him ; with his extended left hand he holds his bow, and seems to be parrying with it."† This beautiful figurine, in which we are tempted to recognise a copy of the Heracles made for the Thasians by Onatas the Æginetan, has a striking resemblance in style to the statues of the pediments of the temple in Ægina. It is certain that it came from the same school, and it shows how the art of working in bronze kept pace in the fifth century B.C. with the art of sculpture.

II.—*Statuettes of later style.*—We shall not undertake to enumerate the most remarkable of the small bronzes of the finished style, owned by our museums and by private collectors. The bronzes from Pompeii and Herculaneum gathered in the Museo Nazionale of Naples, form in themselves a rich museum for study, and reveal to what perfection the Greeks had learned to carry the art of working in bronze. Italo-

---

* De Witte, *Rev. arch.* 1873, plate VI.
† O. Rayet, *Monuments de l'Art antique ;* 1re livraison.

Greek bronze statues, however, of an earlier time, from villages in the Campagna, are still quite rare. Under this head we would specify a work of singular perfection, which sums up in itself all the characteristics of contemporary Greek style in the great schools. This is a statuette representing Aphrodite dressed in Dorian costume, with simple straight folds, wearing a crown decorated with jewels ; the head is charming in its purity, while its exquisitely elegant outlines at once suggest the most finished works of the close of the fifth century B.C. The still hieratic attitude, the majestic fall of the folds of the costume seem to be a sort of concession made to the severe traditions of the fifth century B.C., while the face has all the grace of the succeeding age.

FIG. 134.
BRONZE FROM TARENTUM.

We may ascribe to the art of the fourth century B.C. the beautiful bronze from Tarentum, reproduced in Fig. 134, which seems to represent a military commander addressing his troops. There are few antique bronzes superior to this statuette in the finish of their modelling and in the successful attempt to present individual types.

The art of bronze-working soon followed sculpture in the way opened up by the masters of the new Attic school. The fancy of toreuticians is apparent in charming works of exquisite taste, controlled by the desire for grace and for thoughtful fantasies: drunken Fauns, drinking Heracles, big-bellied Sileni stumbling along under the spell of intoxication, winged Victories; these are favourite subjects with artists working in metal.* This new taste appeared in Græco-Syrian statuettes from Tortosa, dating from the time of the Seleucidæ, among which representations of Aphrodite are very numerous. Græco-Syrian bronzes form the transition from the Hellenic art of the period of independence to the Italo-Greek bronzes, with which the people of Pompeii adorned their dwellings.†

§ 2. OBJECTS FOR ORNAMENT. PLAQUES IN BRONZE REPOUSSÉ.

It is often difficult to determine with precision the use of reliefs in bronze repoussé which form an interesting class among articles in metal. Were they used to decorate furniture? did they belong to

---

* The Antiquarium of Berlin possesses a charming figurine of a Satyr found at Pergamum, contemporary in style with marbles of the Great Altar : Furtwängler, *Satyr aus Pergamon*, 1880.

† The Berlin Museum possesses a very beautiful specimen of the bronzes of the Hellenistic epoch ; it is a group, found at Aphrodisias in the valley of the Mæander, representing Theseus and the Minotaur : A. Conze, *Theseus und Minotauros*, 1878.

pieces of armour? or were these metallic plaques,
often very thin, designed to be sewn upon stuffs or
upon strips of leather? All these hypotheses are
probable, and are justified by the workmanship of
the plaques themselves, which are often perforated
with holes, for the purpose of affixing them to some
other material.*

The Carapanos collection has contributed the
greatest number of specimens of this kind of work.
Amongst them we will mention only a plaque in
repoussé, which doubtless was part of a breastplate. The subject treated is one often represented
on painted vases and on marble votive slabs, viz., the
dispute between Apollo and Heracles for the possession of the Delphic tripod. The two divinities are
treated in the hieratic style preserved by tradition,
but the purity of the lines of the profile and the quiet
grace of the contours show that the archaism is
affected : the work is carefully wrought, and exhibits
the style of the fourth century B.C. If it were necessary to prove what grace and beauty were given by
the Greeks even to industrial articles, no argument
could be more decisive than the citation of another
plaque from Dodona, representing the combat between Pollux and Lynceus. (Fig. 135.) This bas-
relief decorated merely the cheek-guard of a helmet;
and yet the modelling of the flesh parts and
the treatment of the drapery are managed with

* [The Siris Bronzes in the British Museum, which were shoulder-
straps of a Greek cuirass, are among the finest specimens known of
toreutic art.]

the care found in works of larger size. Here can

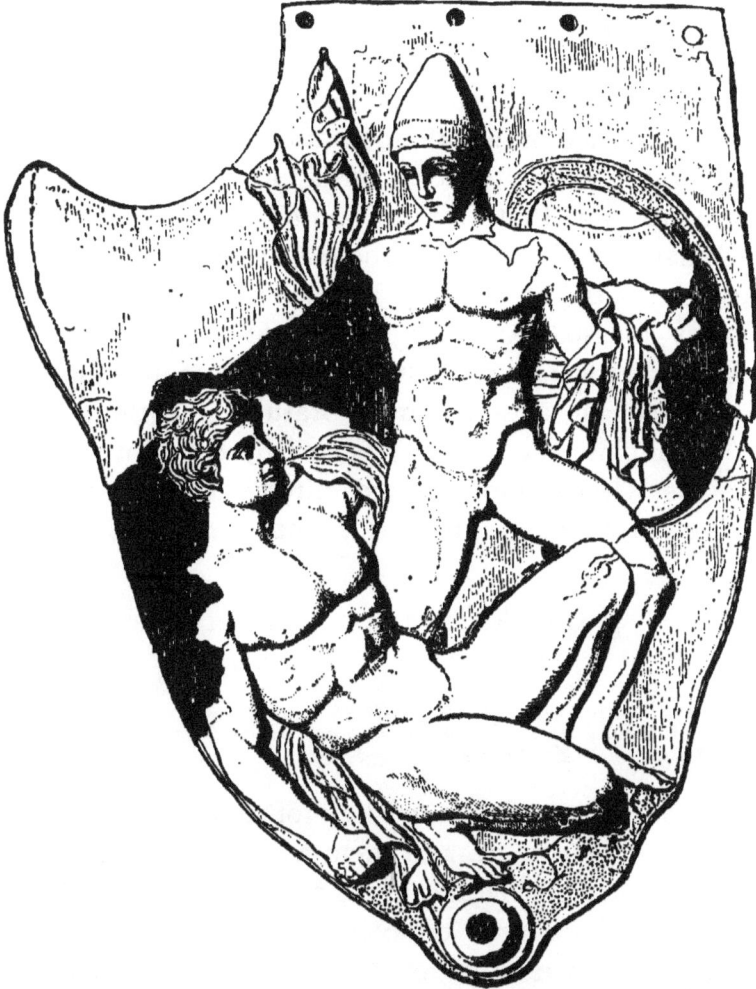

FIG. 135.—COMBAT BETWEEN POLLUX AND LYNCEUS.
(Bronze Relief from Dodona.)

easily be recognised the style of the epoch of
X

Lysippus, and this object takes rank among the most beautiful works of art known to us of the fourth century B.C.

Certain pieces in the Dodona collection, which are very strange in appearance, testify to the taste that originally ruled in the ornamentation of Greek armour. Among the helmet cheek-guards found by Carapanos, there are some that imitate the human face ; these are real masks, and reproduce the features of the warrior's face, the detail of his beard either carefully curled or heavily massed, and the curves of his moustache. The soldier wearing this helmet, and in full panoply, resembled an actual bronze stature. Nothing can better explain the legend reported by Herodotus, that the first Greek hoplites who entered Egypt seemed to the natives actual men of bronze (χάλκεοι ἄνδρες).*

3. ARTICLES OF THE TOILET.—MIRRORS.

DE WITTE : *Les Miroirs chez les Anciens*, 1872 ff.
MYLONAS : 'Ελληνικὰ κάτοπτρα, Athens, 1876.
DUMONT: *Miroirs grecs ornés de figures au trait* (in *Monuments grecs de l'Association des Études grecques*), 1873.

It is to the Greek custom of burying with the dead all that was dear to them when alive that is due the fact that discoveries of tombs, made accidentally or in regular course of excavation, have brought to light a great number of toilette articles : brooches, colour-boxes, mirrors, etc. Among these articles, Greek mirrors form an important class, which has

---

* Herodotus, ii. 152.

only recently attracted the attention of archæologists. Etruscan and Roman mirrors have long
been known in great numbers. Greek mirrors are
infinitely superior in beauty to Etruscan mirrors,
and prove that Hellenic art did not disdain to ornament these accessories in the toilette of women with
most exquisite care.

Greek as well as Etruscan mirrors are generally
rounded in form. According to their technique, they
are divided into two classes :—1. Simple mirrors in the
form of discs, with a carefully polished convex front
surface that reflected the image, and a concave back
ornamented with figures traced with the engraver's
burin. These discs were provided with a handle, in
the form of a statuette with pedestal, which allowed
them either to be held in the hand or to stand upright upon a table. Painted vases with toilette scenes
often represent women holding their mirrors. On a
vase in the British Museum a woman looks at herself in
the mirror, and on another is the inscription : αὐτοψία
("view of herself"). 2. Another form, especially
common in Greece, is that of two mirrors combined
into a sort of case. These mirrors are composed of two
metallic discs, the one enclosed within the other, which
are sometimes held together by a hinge. The upper
disc, or cover, was ornamented on the exterior with
figures in bas-relief, and on the interior was carefully
polished or silvered ; it was this face that reflected
the image. The second disc, forming the body of
the case, was decorated on the interior with figures
engraved with a sharp point. The outlines of the

X 2

figures are often filled in with a thin thread of silver, while the background is gilded.

It will thus be seen that Greek mirrors, in their technique, furnish a threefold subject for study : (1) engraving in outline ; (2) bas-reliefs ; (3) handles in the form of statuettes.

(1) *Mirrors engraved in outline.*—Etruscan artists practised the art of engraving with a burin with great skill, as is proved by numerous mirrors found in Etruria. But in this matter, as in many other matters, the Greeks were their masters. There is no question that the Greeks were initiated into this art by the Orientals,* but it must be admitted that the Greeks carried the art of engraving on bronze with a burin to the highest perfection. This great skill, united with the rarity of the specimens known, gives a high value to the Greek mirrors that have been preserved.† One of the most beautiful examples among engraved mirrors is one representing the hero eponymous of Corinth (Κόρινθος) crowned by a woman (Λευκάς), who personifies the Corinthian colony of Leucadia. The hero, in part naked, is seated in a chair with massive supports ; a mantle covers the lower part of his body, and his torso appears in all the beauty of its vigorous forms ; the head has a calm expression of power. He turns

---

* See metal cups engraved with a burin from Cyprus, *La Patère d'Idalie*, in *Rev. arch.*, 1872.

† On the number of mirrors known, see A. Dumont, *Miroirs grecs*, in *Bull. de Corr. hellén.*, 1877 ; and M. Collignon, *Bull. de Corr. hellén.*, 1884. The first was discovered at Corinth in 1867.

towards Leucas, who, robed in the himation, is about
to crown him. A rosette and designs of marine
flowers and plants in the field of the mirror complete

FIG. 136.—LEUCAS AND CORINTHUS.
(Greek mirror engraved with a burin.)

the composition. The simple taste that controlled
the arrangement of the scene, the perfection in the
drawing, and the boldness of certain details, bear
witness to the finest qualities of Hellenic art. When

we compare this mirror with the most beautiful Etruscan mirrors, we immediately recognise what a distance separates this chaste and elegant style from the heaviness of Etruscan designs. Among other Greek mirrors, one representing the genius of cock-fights (in the Lyons Museum), and one representing dancing girls, hardly yield the palm to this Corinthus mirror. The study of these monuments reveals the perfection of Greek art in a department that was long believed to be the exclusive possession of Etruria.

(2) *Covers of mirrors with bas-reliefs.*—Bronze bas-reliefs decorating the covers of mirrors belong to a class already known ; they have not the advantage of representing a branch of Hellenic industry hitherto ignored. But we must not underrate the artistic value of these reliefs, several of which belong to the best epoch. The bronze relief showing Ganymede borne off by an eagle is a masterpiece of graceful art. (Fig. 137.) Upon another relief a drunken Silenus carries away a Mænad, whose knee rests upon his hip, while an Eros with outstretched wings flies before them. Although the style may be that of the later time, it has preserved all the beautiful delicacy of the best period of art. In general, reliefs on mirrors* exhibit subjects taken from the cyclus of Aphrodite, and from that of Dionysus, divinities of a graceful, sensuous character, whose merry followers, Loves, Pans, and Mænads, inspired the spirit of the artist ; these

---

* They are few in number—hardly more than thirty.

subjects are, furthermore, in perfect keeping with the uses to which such decorated objects were put.

(3) *Feet of Mirrors.*—Since Greek mirrors have

FIG. 137.—GANYMEDE BORNE AWAY BY THE EAGLE OF ZEUS.
(Greek mirror, with relief.)

begun to be studied with greater care, it has been recognised that several statuettes, Græco-Italic in style, already known, served as handles or supports for mirrors. This class of monuments has been rapidly enriched of late. From the point of view of art, these figurines form a very interesting group. The most ancient of them have all the stiffness of archaic art, and remind one of the female figures that adorned the ridge of the Æginetan temple. One of them, belonging to a private collection in Athens, exhibits Aphrodite in her ancient form ; the goddess, closely enveloped in a robe that suggests the outlines of her body, her feet placed together, holds in her hand a dove, while two winged chimæras, with elevated wings, support the disc of the mirror. Quite different is the small figure we reproduce (Fig. 138). The style is already in its perfection, and the erect position of the body, the vertical folds of the dress, in themselves recall the severe style of the preceding epoch ; two winged Loves support the mirror, around the border of which runs an elegant row of ovolos.*

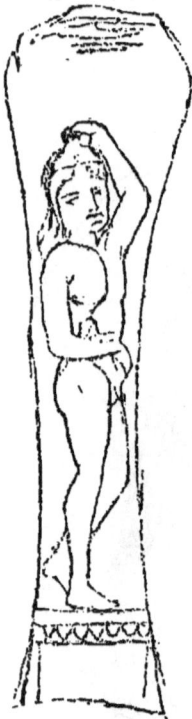

FIG. 138.
HANDLE OF A
STRIGIL.
(Engraved with a
burin. Athens.)

It is impossible here to enumerate all the bronze articles known at present. This is the task for the branch of archæological science known as *museography*. Special

* This figurine was drawn by M. Chaplain for a work, as yet unpublished, by MM. Dumont and Chaplain, to whom our thanks are due for the engraving.

FIG. 139.—FOOT OF GREEK MIRROR.

attention, however, should be drawn to engraved bronze discs, which are few in number, and must not be confounded with the mirrors. The Berlin Museum possesses a beautiful specimen, representing an ephebus exercising with *halteres* or weights used in jumping. More commonly these discs are votive offerings. The process of engraving seems to have been applied to other objects also, as plaques used to decorate furniture, or the handles of strigils, articles employed in the palæstræ and gymnasia, with which athletes scraped their bodies when covered with sweat and oil. The handle figured above (Fig. 138) is in the Varvakeion Museum, in Athens.

The study of small bronzes still promises interesting discoveries. Such objects show that nothing can be disdained by archæology. The examination of these tiny monuments contributes to the development of skill in discernment or artistic tact, to the education of the eye, without which archæological studies can have no existence. At the same time, nothing can make us better appreciate how thoroughly united were all the Greek arts; the Hellenes could not permit objects of daily use, products of the humblest art, to be without a certain beauty for which they felt an instinctive need.

CHAPTER II.

JEWELLERY.

DE LASTEYRIE: *L'Orfévrerie depuis les Temps anciens jusqu'à nos Jours*, 1875.
BILLING : *The Science of Gems, Jewels, etc.*, 1867.
CLÉMENT : *Catalogue des Bijoux du Musée Napoléon III.*, 1862.
ARNETH : *Die antike Gold- und Silbermonumente des Antiken-Cabinet in Wien*, 1850.
SAGLIO : *Cælatura*, inDaremberg and Saglio's *Dictionnaire des Antiquités grecques et romaines*, 1873 ff.

WORK in precious metals constituted an important branch of Toreutics, and at the outset the Greeks did not separate these two arts ; in the Homeric age the worker in bronze (χαλκεύς) is at the same time the goldsmith (χρυσοχόος).* Greece, for a long time poor in the precious metals, attributed to the art of working in gold and in bronze a legendary origin. The Cabeiri and the Dactyli were the first blacksmiths, and the Telchines were the first goldsmiths. We have already remarked that these legends most commonly disguise the fact of borrowing from the Asiatic Orient. The part taken by the East in the initiation of the Greeks into the processes of these minor but dainty arts is evident ; jewels of gold and silver, and vases of precious metals, were pre-eminently objects of commerce, introduced by the Phœnician traders. The discoveries at Mycenæ leave no doubt on this point.

* Homer, *Odyss.* iii. 425—432.

In the Homeric age the articles in metal most in favour came from abroad; the highest praise that Homer can bestow upon them is to say that they were made at Cyprus or at Sidon. Agamemnon's breastplate and chariot were a present from the King of Cyprus;* the silver crater given as a prize by Achilles at the funeral games in honour of Patroclus is the work of Sidonian artists.† The monuments that best enable us to understand Homeric descriptions are pieces of jewellery found in Cyprus and in Rhodes, and even in Italy.‡ These cups, which are of engraved silver, worked in repoussé, exhibit warlike or religious scenes, arranged in zones, and give us an accurate idea of the precious vases imported by Phœnician sailors into Greece, which furnish the principal elements of the Homeric descriptions.

Furthermore, jewels of the earliest style, found in Greek countries, reveal distinctly the influence exercised by the East upon objects of luxury and ornament of the epoch of the eighth century B.C. Not to multiply examples, we will mention an interesting band of stamped gold found at Athens,§ which shows a procession of animals distinctly suggesting, both in arrangement and in style, the zones

---

* Homer, *Iliad* xi. 19. † Homer, *Iliad* xxiii. 741.

‡ See the cups from Cittium and Larnaka in the Museum of the Louvre, and especially the beautiful patera from Amathus, published by Colonna-Ceccaldi, *Rev. arch.*, 1876. [*Cf.* Helbig, *Das Homerische Epos*, 1884.]

§ Now in the Louvre. This museum possesses gold ornaments, studs, or earrings, decorated with human heads, where the hair is dressed in Egyptian style. These jewels were found at Megara.

of Asiatic animals. The most important group of jewels of this period is composed of those discovered at Rhodes by Salzmann, in the necropolis of Cameirus. The Oriental style characterising them is apparent in the stamped plaques of very pale gold, that belong to a necklace. They exhibit two types alternately, in the one a Centaur with human feet and Egyptian head-dress, represented in his most ancient form, in the other an Asiatic goddess, either the Persian Artemis or Anaïtis, holding a lion or a panther by each hand.

With decorative types, the Greeks also received from the East a knowledge of technical processes ; thus the art of working in precious metals, as well as in other metals, was developed in Asiatic Greece. The toreuticians of Samos and of Chios were goldsmiths as well, and they quickly attained such skill in workmanship that the East borrowed from them in turn. Henceforth Greek artists lavished on the execution of objects of luxury and ornament such exquisite taste and such perfection in technique, that even the Etruscans, consummate masters of the art of working in precious metals, were constrained to become their imitators.

Jewels found in Greece proper are quite rare ; one of the richest collections of Greek jewels, that of the Hermitage in St. Petersburg, comes from the ancient Panticapæum (now Kertch) in the Crimea.*

* These jewels are published in the *Antiquités du Bosphore Cimmérien*, 1854, and in the *Comptes rendus de la Commission archéologique de Saint-Pétersbourg*. [See C. T. Newton, *Greek Art in the*

How did works of art, evidently of Greek origin,
penetrate to a country so remote? The fact need
not surprise us, if it be remembered that in the fifth
and fourth centuries B.C. the fame of Greek artists had
extended so far that even princes of the Cimmerian
Bosphorus secured their services either at Athens or
in their own country. Another locality still rich in
Greek jewels is Etruria : the Etruscan tomb-cities of
Vulci and of Cære furnished the materials of the
collection in the Gregorian Museum in Rome, and of
the Campana collection, the property of the Louvre.
In spite of their Etruscan origin, these articles are
none the less valuable in the history of the art of
working in gold among the Greeks. They belong to
a period when ancient Etruscan art, which was Asiatic
in origin, had given place to an art derived from
Greece. At this time the relations between these
countries were very intimate, and Hellenic taste ruled
in these dainty pieces of work, of which the carefulness of detail was in perfect keeping with the
peculiarities of Etruscan genius.

The technique of Greek jewels still presents many
problems that demand careful study. In our day
Roman jewellers, the MM. Castellani, have attempted
to solve them by practical experiments, and by imitations executed with remarkable skill.* And yet it

---

Kimmerian Bosporos, in his *Essays on Art and Archæology*, 1880,
p. 373 *ff*.]

   * Castellani, *Communication faite à l'Académie des Insc. et Belles-
Lettres*, 20th December, 1860, and *Della oriticeria Italiana*, Rome,
1872.

must be admitted that on certain points the secrets
of these ancient goldsmiths have not been discovered
and disclosed ; it is still a matter for inquiry as to
how Greek and Etruscan artists gave their works
such an inimitable beauty and finish. Granulation, a
kind of decoration that consists of covering the surface of gold-leaf with minute and almost invisible
bosses of gold, a process employed in the majority
of Etrusco-Greek jewels, is one of the secrets that
modern art despairs of discovering.

The guiding principle of Greek artists seems to
have been the following: to regard workmanship as of
far greater importance than the materials employed.
Thus it was not the intrinsic worth of precious metals
that gave to Greek jewels their high value ; it was
the exquisite work upon the metal, and the fancy
exercised in the ornamentation. The elements of ornamentation are taken from nature ; they include fruits,
flowers, and foliage, with which is joined imitation
of the human body. All these motives are treated
with an art that is rich in resources, the sole law in
which is the artist's fancy. At the same time, the
use to which the jewels were to be put imposed certain limitations upon the artist ; thus fancy was freer
in the decoration of pendants of earrings and of necklaces, than in that of bracelets or crowns, where art
was more severe.

Before specifying certain types adopted by Greek
jewellers, it is necessary to distinguish, among ancient
jewels, those that were intended to be worn, and
those that were used in connection with the dead.

The latter were merely objects for display, placed upon the dead on the day of the funeral, and buried with him. It was not uncommon to save real jewels by providing substitutes for them in this way. Such are the crowns found in many Greek tombs; * they imitated the foliage of the oak and of the laurel, etc.; but the extreme thinness of the gold leaves, beaten where they had been cut, shows clearly that they were made only for transient use. It is hardly necessary to remark that the jeweller's art reserved its best resources and its highest skill for the ornaments destined to be permanently worn.

It would be difficult to pass in review all the articles designed for ornaments in use among women, upon which the goldsmith exercised his fancy; individual taste and caprice, as well as fashion, prevailed, and thus was created an infinite diversity of types. We will mention only a few examples. One of the most beautiful Etrusco-Greek jewels in the Louvre is a woman's diadem, or *stephane*, where beads of glass-paste and enamelled palm leaves are combined with ornaments of engraved gold." †　This diadem is in imitation of a wreath composed of daisies and other smaller flowers, with which are combined bunches of foliage of exquisite work-

* The Museum of Varvakeion, in Athens, possesses golden crowns of this class. Such may also be seen in the Louvre, in the Hermitage Museum at St. Petersburg, and in the Gregorian Museum in the Vatican. Ancient writers make reference to this custom. Menander (*apud* Stobaeus, *Flor.* cxiii. 2) speaks of a richly adorned corpse: πολυτελὴς νεκρός.

† *Bijoux du Musée Napoléon III.*, No. 1.

manship. This ornament is very different from a
superb golden crown found at Armento, now the
property of the Antiquarium at Munich. The latter
is composed of branches of oak-leaves, with which
intertwine garlands of flowers, while in the upper
portion winged figures stand in the midst of the leaf-
work. The inscription cut upon the side of one of
these figurines indicates the votive character of this
work of art: "Creithonius dedicated this crown."
Perhaps it formed part of the treasure of some
temple.

Necklaces were often composed of several rows of
small chains, with pendants; the central pendant,
largest in size, showed the most careful workmanship,
and represented a flower, or the head of a divinity.
The Museum of the Hermitage, in St. Petersburg, is
very rich in jewels of this class, which were found in
the Crimea. We mention especially a magnificent
pendant, from the tomb of a priestess of Demeter.
Small chains and tassels of gold are suspended from
an engraved plaque that represents a Nereid carrying
the greaves of Achilles' armour. It was, however,
chiefly in the execution of the pendants of earrings
that goldsmiths showed their extraordinary fertility
of invention. At the same time the motive is very
simple—an engraved button, surmounting a smaller
figure worked in repoussé, with a skilful combination
of chains and figures, some of which are placed in
groups. One of the most wonderful instances of
this skill of workmanship is a pair of earring pendants found in a tomb in Bolsena; here, on the

Y

tiniest scale, the artist has reproduced the chariot of the Sun, driven by the god, accompanied by winged Victories. Above this is a sort of cupola, from which hang small chains, terminated by palm-leaf ornaments and tiny amphoræ. At the sight of this marvel of technical skill we may without difficulty form some conception of the masterpieces of delicate work ascribed to Greek toreuticians, such as Callicrates of Lacedæmon, and Myrmicides of Athens. It was always the same motive—a small iron quadriga that a fly might cover with its wings.

Bracelets and fibulæ, or pins for the peplus, are numerous in our museums. Ordinarily the work is more simple than in the earrings ; the bracelet is commonly composed of large engraved circular discs, or of plaques in repoussé, united by bands and furnished with a clasp. The one here figured (Fig. 140), which is of the Roman epoch, is composed of Mitylenæan hectæ, mounted with small garnets. In Greek tombs are also found gold and silver plaques of repoussé work, that must have been used to decorate garments. Such is the plaque (Fig. 141) found near Athens, giving a very pretty specimen of the subjects that artists liked to treat: a young girl is

FIG. 140.—GOLD BRACELET.
(Found in Epirus.)

weighing two Loves in a balance—an exquisite illus-
tration of some epigrams in the Anthology, where
poets express with keen analysis the subtlest emotions
of the soul. The mythology of love inspired these
artists in decorating articles for ornament, and in
making them conform to the
refined taste of the more cultivated classes of society.

FIG. 141.
SILVER PLAQUE.
(Found in Attica.)

To the art of chasing are
due those large pieces of jewellery, the use of which was very
common in Greece in the
Macedonian epoch. Under the
successors of Alexander, Greek
life lost its original simplicity,
and the luxury displayed in
the courts of these rulers con-
tributed rapidly to the growth of this branch of
industrial art.* Thus the artists mentioned by
Pliny as masters in the art of engraving are for
the most part contemporaries of the Ptolemys
or of the Seleucidæ. Mentor, Acragas, Boëthus, are
the most brilliant among these toreuticians, who,
designated by the term "minor artists" (μικροτέχναι),
exercised their talents in decorating with reliefs vases
of precious metal. We can hardly estimate their
talent without the aid of ancient writers ; silver vases
of the best Greek period are extremely rare. The

---

* See in Athenæus (v. 29-30) the description of a spectacle
given by Ptolemy Philadelphus, where were carried vases with Bacchic
subjects and theatrical masks, chased in high relief.

Y 2

Munich vase, representing captive Trojans, and the
vase at the Hermitage (found at Nicopol), on which
are figured Scythians in their national costume, reveal
a pure Greek style, but are probably of a later
date.   In the Roman epoch the taste for vases of
gold and silver became general, but the style suffered
marked degeneration.   There was a profusion of
ornament, relief became of exaggerated height, and
everything was sacrificed to an evident striving after
richness of effect.   Among the specimens of ancient
silver ware that have come down to us, there are very
few in which purity of style and moderation in the
use of ornament bear witness that they are copies of
Greek originals.*

* For example, a cup found at Porto-d'Anzio, known as the Corsini
Cup.   See Michaëlis, *Das corsinische Silbergefäss*, 1859.

GENERAL INDEX.

—◦◇◦—

GREEK INDEX.

GREEK INSCRIPTIONS QUOTED.

www.ingramcontent.com/pod-product-compliance
Lightning Source LLC
Chambersburg PA
CBHW021354210326
41599CB00011B/876